SENSATIONAL SYSTEMS

10 Techniques for Delivering Successful Systems

Geoff Miles

Sensational Systems: 10 techniques for delivering successful systems
Miles Geoff
www.sensationalsystemsbook.com

Copyright © 2018 by Geoffrey Miles

ISBN-13: 978-1727733341
ISBN-10: 1727733347

All rights reserved, including the right to reproduce this book or portions thereof in any form whatsoever. No part of this publication may be reproduced, stored in a retrieval system or transmitted in any form or by any means, electronic, mechanical, photocopying, recording, scanning or otherwise.

It is illegal to copy this book, post it to a website, or distribute it by any other means without permission from the publisher or author.

Limits of Liability and Disclaimer of Warranty:
The author and publisher shall not be liable for your misuse of this material. This book is strictly for informational and educational purposes.

Warning – Disclaimer
The purpose of this book is to educate and entertain. The author and/or publisher do not guarantee that anyone following these techniques, suggestions, tips, ideas, or strategies will become successful. The author and/or publisher shall have neither liability nor responsibility to anyone with respect to any loss or damage caused, or alleged to be caused, directly or indirectly by the information contained in this book.

Published by:
10-10-10 Publishing
Markham, Ontario, Canada

First 10-10-10 Publishing paperback edition September 2018

Table of Contents

Dedication	vii
Foreword	ix
Acknowledgements	xi

Chapter 1: What Do We Mean by a System? — 1
- What Do You Want to Achieve with Your System? — 1
- How Will You Know That You Have Succeeded? — 2
- Establishing and Reviewing Your Success Criteria — 3
- Your Objectives Divided Into Manageable Chunks — 6
- Creating a Picture of Your System — 7
- Measuring What You Want to Manage — 9

Chapter 2: Defining the System — 11
- System Boundaries — 11
- Defining Your Inputs — 13
- Defining Your Outputs — 14
- Identifying External Influences — 16
- Environmental Effects — 16

Chapter 3: System Stakeholders — 19
- Who Wants Your System? — 19
- Users, Owners, and Maintainers — 21
- Who Will Benefit From It and Who Will Pay For It? — 25
- Internal and External Stakeholders — 27
- Stakeholder Aspirations — 29

Chapter 4: What Could Possibly Go Wrong?   **33**
- Consider the Risks   33
- Likelihood of Risks Occurring and the Impact if They Did   35
- Risk Ranking and Tolerance to Them   37
- Risk Management Process   39
- Risk Mitigation and Ownership   44
- Compliance With requirements   46

Chapter 5: Systems Within Systems   **53**
- Simple or Complex   53
- Interdependencies   55
- Eliminating Complexity and Interdependence   56
- Establishing Interfaces and Responsibilities   61
- Helping Others to Understand Your System   63

Chapter 6: Building Your System   **67**
- System Components and Build Sequence   67
- Defining the Work Breakdown Structure   70
- Lead Times and Priorities   75
- Resources and Logistics   79
- Proven or Novel Elements   86

Chapter 7: Testing and Commissioning   **91**
- Measurements   91
- Test Plans and Schedules   94
- Stakeholder Involvement   97
- Simulating Failure and Recovery   99
- Getting Acceptable Results   102

Chapter 8: System Assurance   **105**
- Creating a V Life-cycle Model for Your System   105
- The Assurance Plan   108
- Collecting Evidence   114
- Progressive Assurance   117
- Stage Sign-off   119

Chapter 9: Operation and Maintenance **125**
- The Training Plan 125
- System Operating Regime 128
- System Maintenance Regime 132
- Operation and Maintenance Documentation 135
- Asset Management 138

Chapter 10: Usign Your System **145**
- Performance 145
- Upgrades and Obsolescence 147
- Stakeholder Satisfaction 149
- Assessing Whole Life Benefits 152
- Learning From Experience 154
- Decommissioning 156
- Enjoying Success 158

About the Author 161
Testimonials 165

Dedication

I would like to dedicate this book to my loving wife, Diane, who has stood by me for more than forty years during my successful and sometimes not so successful adventures in life. She has been a wonderful mother to our two terrific children, Katherine and Benjamin, and a supportive grandmother to three grandchildren, Kia, Evie, and Austin.

I would also like to thank my parents, Leslie and Betty, who did a fantastic job in bringing me into the world and, for seventeen years, showed me right from wrong and taught me to have integrity in all that I did. I'm not sure that they followed a Sensational System, but they must have done some things right because they did them successfully, not only for me but also for my two sisters, Janet and Barbara, and my three brothers, Raymond, William, and Robin.

Foreword

You may find that systems dominate your regulated and crowded life. Many systems now contain software which, often as not, fails to perform reliably with the blame falling on you, who developed the system in first place. Systems can be soft or hard and they can be simple or complex. This book will show you not only the differences but also the similarities and, most importantly, will show you the basic principles that can be applied to good effect to any system.

You may note that this book does not have the word 'Engineering' in the title. Whilst systems engineering is a relatively modern discipline, Sensational Systems is about more than engineering of systems; it will give you insights into methods and techniques for delivering success in a whole range of processes which could be considered as systems. 'How to get to work on time' is a system you may find enlightening.

Of course, many systems are about engineering, and if you put Systems Engineering into the Google search box you'll find at least fourteen books at the top of the first page. In fact, more than one hundred books on systems engineering have been published since 1950. Interestingly, more than half of them were published in the 1990's; it was at that time that US academics started recognising systems engineering as a professional discipline and early academic papers turned into books.

It is still true that many of the currently available books on systems appear to be heavy on theory and rigorous processes, and this is where Sensational Systems, authored by Geoff Miles, is refreshingly different. Geoff uses his extensive experience in dealing with complex issues and problems in the rail industry and providing solutions which are simple to comprehend.

I am sure that you will find this book helpful in determining a clear path through the jungle of complexity in many aspects of your life. You will want to keep it in your library as a resource and as a powerful tool.

Raymond Aaron
New York Times Bestselling Author

Acknowledgements

I am privileged to be a Chartered Engineer and a member of the Railway Division of the Institution of Mechanical Engineers, following in the footsteps of George Stephenson (the first president and Father of the Railways), William Armstrong (whose house, Cragside, was the first in the world to use hydroelectricity), Arthur Hartley (who helped develop the Pluto Pipeline to supply the Allies during World War II), Bernard Crossland (who carried out the Kings Cross fire investigation) and Pamela Liversidge (the first female president). Whilst my own career may not have been as illustrious as many of the Institution's past presidents, it has been coloured by a number of mentors, colleagues, and people of influence whose contribution to shaping my views on successful systems are acknowledged below.

The staff at Rycotewood College Oxford, especially the vice principal, **Mr Coles**, 1969 to 1971, got my engineering career launched with a National Diploma, and this was followed by a Bachelor of Science degree awarded by the University of Aston, in Birmingham. I was a student sponsored by Metro Cammell, and I am indebted to my mentor of 13 years, **Tony Thomas,** who was Training Manager, Personnel Manager, and later became Works Manager, at the company in Washwood Heath. The

Commissioning Manager at Metro Cammell, **Jim Fallon,** showed me how to drink Irish whiskey, get complex train systems working reliably, and to have fun doing it.

As the Commissioning Engineer in Hong Kong, for the initial fleet of trains for the Mass Transit Railway, I am grateful to **Edmund Leung** for all his help and support, to **Nigel Twort** for air conditioning systems know-how, and to **Chris Llewellyn** and **Justin Stopford** for electrical systems problem resolution, without electrocuting myself. The **Kowloon rugby team,** where I was loose head prop, 1976 to 1979, provided entertaining systems knowledge within the *clubhouse* of an Australian bar off the Nathan Road.

With the dawn of commercially available computers in the 1980s, I acknowledge the opportunities presented by **John Floyd** and **Brian Ransom,** at Strachan & Henshaw, for me to learn about and implement computerised production control systems. These modular systems were used to efficiently produce weapon handling systems for the fleet of Trafalgar and Trident class submarines for the Royal Navy. They also enabled cost effective production of the nuclear fuel handling systems at Heysham and Torness Advanced Gas Reactor power stations, and the nuclear fuel storage and reprocessing plants at Sellafield and Aldermaston.

In 1989, I joined **Mike Nichols** as his 34th employee at The Nichols Group. I readily acknowledge Mike's personal support and guidance in delivering sensational systems throughout my twenty-seven years as a management

Acknowledgements

consultant. His company mission and vision were to delight clients, exceed their expectations, and have fun doing it. I hope to achieve this mission and vision for readers of this book. Mike's daughters, **Kathryn** and **Frances Nichols**, thankfully have continued to direct the business in the same style following their father's untimely passing. I wholeheartedly acknowledge the friendship, mentoring, and support given by colleagues, past and present at The Nichols Group, including **Peter Hansford, Ernie Torbet, Bill McElroy, Martin Buck, David Waboso, Stuart Westgate, Tony Haworth, Colin Britt, Stephen Jones, Graham Tillett, Colin Jurd, Jacqueline Beall, Emma Brookes, Simon Web, Tony Salmon, Jan Woods, Pat Stirling, Chris Ctori, Alistair Godbold, David Hicks, Paul Johnston, Aman Sharma, Kerry Bangle, Paul McAleer, Andrew Elliott, Nigel Dumbell, Rob Little, Jonathan Holland, Heather Elliott, Paul Bishop, Colin Thomas, Marie Gilmour, Louise Toppin, Trevor Giddings, Mark Jones, Tim Bowling, Louise Pengelly, Terry Ginever, Colin Dash, Hilary Moules, Tony Calthorpe, Mike Stephens, David Cawthra, Carl Chouler, Danny Forker, David Grey, John McLaughlin, Gary Jones, Vanessa Jackson, Kevin Thorpe, Tracy Samwell, Carol Lee, Phil Morris, Jeremy Richold, Simon Young, Mark Thurston, Tim Brooking, Katie Kendrick, Ralph Goldney, David Yass, Jeff Done, Christine Stephens, Jonathan Jong, Sarah Thurston, Ros Marks, Wilson Manson, Richard Preston, Phill Stanton, Mo Williams, Nermeen Latif, Malcolm Maskill, Kevin Cooper, Jonathan Morris,** and **Tony Fletcher.**

I'd like to acknowledge professional colleagues from railway signalling system backgrounds; their expertise continues to ensure the safety of many thousands of people who travel by train each year. These include **Chris Heavens, Chris Binns, Colin White, Laurie Kent, Steve Massey, and Phil Clayton;** their contribution to my systems knowledge has been invaluable.

The multi-national consultancy, Mott MacDonald, is acknowledged for providing assignments and professionalism for me, particularly in the systems market place. Colleagues include: **Paul Holder, Jim Baker, Chris Bryce, Adrian Featherstone, Russell Smith, Rob Evans, Dara Sarhangian, Dick Dumulo, Matteo Bruni, Shabaz Younus, John Bowes, David Holmes, Wayne Hsu, Michael Curthoys, and Jasmine Purushotham.** Special thanks go to **David Quastel** and **Massimo Santi** for their systems engineering guidance and knowledge.

I acknowledge the valuable work in asset management systems done by colleagues at Atkins consultancy, including **Navil Shetty, Carl Waring,** and **Dr Bernard Rochard**.

London Underground, one of the oldest, largest, and busiest metro systems in the world, has operational and control systems that are world class, and which continue to operate safely with ever increasing numbers of passengers, and train headways of around 90 seconds. I'm proud to have been part of the continuing development of that system for many years, and acknowledge many

Acknowledgements

colleagues past and present there, including **Simon Dean, Richard Miller, Mike Hurn, Bob Mitchell, Gordon Torp-Petersen, Jim Moriaty, Joe Cosgrave, Mel Gardner, John Downes, Graham Neil, and Malcom Dobell.**

I would like to think that my contribution to the concept, initial design, obtaining powers for construction as the first major project to use the Transport and Works Act, and the many systems interfaces of the rail system (which was first known as the East London line, then Outer Circle line for London, and later, Orbirail) which ended up becoming the London Overground, has played its part in enhancing the quality of life for the ever-growing number of London citizens. I acknowledge the help given by colleagues in this venture, which include **Mike Kilby, Brian O'Connor, Duncan Murray,** and **Rashmi Devri**.

Environmental systems and their impact on people (sensitive receptors in the jargon) have shaped my knowledge and appreciation of what an important area of systems this is. I acknowledge the support and guidance of **Mike Adams** in this area.

More than 10% of my working life has been directed towards the Crossrail project, now known as the Elizabeth Line in London. I have been privileged to work with many true professionals in Canary Wharf, and at Old Oak Common depot in West London. I do not hesitate to acknowledge these colleagues, including: **Phil Hinde, Phil Clarke, Andy Lees, Paul Richardson, Chris Nash, Anthony Hartley, Charles Devereux, Fred Drury, Ed**

Hamlyn, Martin Treacy, James Mendis, Dave Sherrin Martin Stuckey, Matiur Choudray, Graeme Overall, Jon Jarrett, Jignesh Patel and **Dave Brignell.**

Creative styling, branding, and graphic design are part of visual systems that influence our lives on a daily basis. Both able-bodied people and people of reduced mobility, partially sighted or blind, all need to use public transport confidently and safely. I have had the benefit of working with one of the gurus in this area of systems, and acknowledge the guidance given by **Siep Wijsenbeek,** of Design Triangle, in a number of the systems I have been involved in.

Similarly, ergonomic design and the ability to use manmachine interfaces for long periods, in a safe and user-friendly way, is so important when conceiving systems and getting them placed in a working environment. I have had the opportunity of working with **Gary Davis** on several systems jobs, and I acknowledge his expertise in this area.

Wayfinding is a system that we might not think about very often, until we have difficulty finding the ear, nose, and throat department at our local hospital, or the nearest public toilet when we are *caught short*. A successful architect in this field, and indeed in the field of excellent public space design, is **Maurice Green,** of Scott Brownrigg, and I value his friendship and support for my systems efforts.

Acknowledgements

My friends and colleagues at Bombardier have provided many systems insights over nearly 30 years of association. I'd particularly like to acknowledge the support given by **Tim Mason, Peter Doolin, Mariena Somasundaram,** and **Ian Forman.**

For those who travelled on the West Coast route in the 1990s, you may remember the Mark 2F coach fleet and grimace. At the time of British Rail privatisation, these coaches formed trains serving the intercity prestige routes. Their ride quality was poor due to lack of maintenance; the air conditioning systems were broken more often than not, due to power supply failure; and the body structure around the toilet cubicles had rusted to the point where the floor flapped up and down as the train travelled at 90 miles per hour. I was fortunate enough to be awarded the project manager's job for sorting out these problems, was given sufficient funds to do it successfully and, importantly, was able to work with a team of exceptional colleagues, including **Vic Young, Pete Jones,** and **Eddie Knorn.** I heartily acknowledge their influence in delivering sensational systems for West Coast travellers before the arrival of Virgin's Pendolino fleet.

British Rail became Railtrack, in 1995, which in turn became Network Rail, after the tragic disaster at Hatfield, in 2001. I have worked with many people responsible for the UK national rail infrastructure systems throughout my career, and here I'd like to acknowledge just a few of those: **Stella Bye, Danny Trup, James Greatbanks, Natasha Quraishy,** and **Mike Dyson.** And in this acknowledgement, I also include all those dedicated railway systems people who continue each day to make our railway work successfully.

I must mention the team at Ilford depot, where infrastructure, signalling, and operational systems were dragged into the 21st century, with the help of myself and a team of stalwart die-hards. These included **Iain Warner, Dave Cooke, Helena Lachowycz,** and **Gary Stokes.**

Finally, with apologies to all those acquaintances and colleagues who may be offended that they have not been named in this acknowledgement (consider yourself here in spirit if not in print), I'd like to acknowledge a number of individuals who have provided me with friendship, help, and advice, for life in general, and in my quest for sensational systems in particular: **Robert Assirati CBE, Glen Jackson, Eddie Turnock, Mick Grace, Don Tibbs, Glen Snowdon, Brian Johnston, Paul Johnson, Bill Tait, David Greenway,** and particularly, **Paul Dawkins,** whose attitude and approach to living provides an enviable model of what a successful system should look like.

Chapter 1

What Do We Mean by a System?

"We can allow satellites, planets, suns, universe, nay whole systems of universes, to be governed by laws, but the smallest insect, we wish to be created at once by special act."
– Charles Darwin

This chapter creates an understanding that a *system* can be used to deal with many different aspects of life, business, leisure activities, the universe, and everything.

What Do You Want to Achieve With Your System?

The important thing is that a system is a means to an end; a way of achieving something.

A system is an entity with more than one element in it. It can be hard, like a lawn mower is a way of getting your grass trimmed without too much physical effort, or soft, like a computer-assisted way of compiling information into your business management methodology.

The very first thing you need to do is to establish what you want to achieve. Think it through and write it down. If necessary, test your achievement statement on colleagues, friends, and family. Make sure that you have clarity on what it is you want to do.

Having clearly stated what it is that you want to do, you will need to generate a system which will deliver your desired entity.

Almost anything can be considered as *a system*. For example, your washing machine, your car, or your house. But these items, and all those like them, contain systems within them, and are themselves part of larger systems. Your car has an engine, which has a fuel management system within it, and the complete car can be considered as part of a larger transport system.

The important thing is to put a boundary around the system that you want to deal with. Chapter 2 looks at this need in more detail.

How Will You Know That You Have Succeeded?

You may be reading this book because you are curious about systems in general, or you may have had some experience in designing systems, or you may even have been a user of systems designed and built by others—whatever your reasons, the ultimate goal is that you want to be successful.

What Do We Mean by a System?

Now, successful is an odd word because, if you think about it, it can mean many different things to many different people. Let's say that you want to be successful in life. It could mean success in a relationship or in a career, or having money in the bank, and so on. The values that you put on these attributes will be different to mine, and taken together as a whole into an individual life, will be different to everyone else.

It is the same with systems. Take the example of a system to get to the top of a mountain. Easy, you might say; you will have succeeded when you stand on the top of the mountain. You will know you are there because you cannot go any higher. However, you may want to do it in the *fastest ever* time; she might want to do it while assisting a mobility-impaired colleague, or while blind-folded, or some other constraining criteria. The point is that defining success demands that we acknowledge not only the overall goal (getting to the top of the mountain) but also the associated requirements to be met as part of the delivery of our successful system (getting to the top in a world record beating time).

The important thing to understand is that you should write down your definition of success so that you will have sight of where you are heading, and will be able to determine when your system has delivered it.

Establishing and Reviewing Your Success Criteria

Your definition of success will generally come with constraints in some form or another. These can be considered as success criteria or requirements.

You want to get to the top of the mountain—which one? Is it the highest in the country, in the continent, or in the world? Is it just the one, or all the peaks over 3000m in the United Kingdom? You need to be clear on what you are setting out to achieve with your system.

In 1981, George T. Doran wrote an article about goal setting, using SMART objectives. Peter Drucker and other leaders in management techniques have embodied SMART objectives into many of their teachings and learnings. For systems success criteria, it may be helpful to think of SMART as being the mnemonic acronym meaning:

Specific
An identifiable area of the system which has a specific requirement associated with it

Measurable
A part or area of the system which is able to be measured, or at least provide an indicator of progress towards a bigger objective

Achievable
Something that can be attained as an identifiable element of the system. Also, something that can be assigned as the responsibility of an individual or team

Realistic
Identify what results can be realistically achieved given the resources available, and is relevant to the system

What Do We Mean by a System?

Time-related
Specify when the required result or output can be achieved by the system

George T. Doran's notion has been modified over the years with similar but different meanings assigned to each of the letters, but generally with the common aim of successfully setting and delivering goals and objectives in creating something in life.

Others have usefully added ER, to become SMARTER. E is to Evaluate what has been produced, then R to Review the output to be able to improve or enhance what has been done for the future. You may find applying this thinking helpful in achieving even more success with your system delivery!

An example of establishing success criteria for your system is to consider a system for landing aircraft safely at a new airport. A whole range of variables must be specified for landing system designers to do their work: the topography of the area around the airport and on the approach to the runway; the prevailing wind direction and the maximum wind speed in which the air traffic controller will allow a plane to land there; permissible fog density at what altitude before a landing is aborted; and so on. Check whether these are specific enough, if they can be measured, etc.

The question of time is worth considering in more detail. It is largely a matter of scale. Let's say that you need to be at work by 8am, and you want to eat a soft-boiled egg for breakfast. At what time should you set your alarm clock?

The next section helps you with the answer.

Your Objectives Divided Into Manageable Chunks

Your objective is to get to work by 8am, having eaten a soft-boiled egg. Your starting point is where you are having a wonderful dream, snuggled up in your duvet. You need to think through what needs to be done, and in which order; i.e. the steps in the process to achieve your objective from your starting point (this is your system) need to be understood.

In this simple example, the steps might be to wake up, shower, get dressed, make coffee, boil an egg, eat it, clean up the kitchen, brush teeth, leave the house, drive the car to work, park, and arrive at your desk.

All these are very simple steps, which we can think of as our system, but let's look at boiling the egg—how is that achieved? We need an egg, obviously, and water to boil it in, a pan to put the water in, a means of heating the water, and something to time the boiling process (because we want it only soft-boiled, so no longer than three minutes please!). We need an egg-cup, a spoon, and a knife. We need toast, and everything that goes with that.

And so on. In reality, then, we can construct our mission of getting to work on time as a complex system, which is comprised of a number of sub-systems. If we *chunk down* our complex system into manageable parts, we will have a better means of controlling and measuring what needs to be done, have more chance of being realistic in an estimate

What Do We Mean by a System?

of the time each step will take, and recognise when each step is complete and we are able to move on to the next element.

You will find that this basic strategy for breaking any complex system down into manageable chunks of activity will work equally well for an aircraft landing system and, for getting to work on time.

Creating a Picture of Your System

We all know the old adage that a picture speaks a thousand words, and a picture of your system will do that for you. Inevitably, you will need to explain what you are setting out to achieve, and the process for how you hope to achieve it—a picture will give you this.

A small bit of historical context helps here. Systems engineering was identified as a discipline by the United States Department of Defence, in the 1980s. This led to a number of papers being written by academics, which looked at various attributes of systems. They tried to identify and evaluate those that delivered success, and those that were missing from less successful ones.

A number of those papers concentrated on the processes needed to deliver systems.

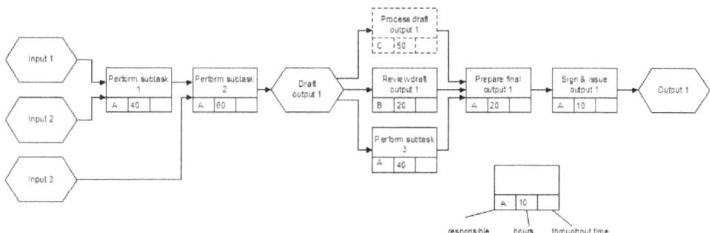

These system processes helped to give clarity to what needed to be done, but they must not be confused with how the system is structured—the System Breakdown Structure. An example is shown below.

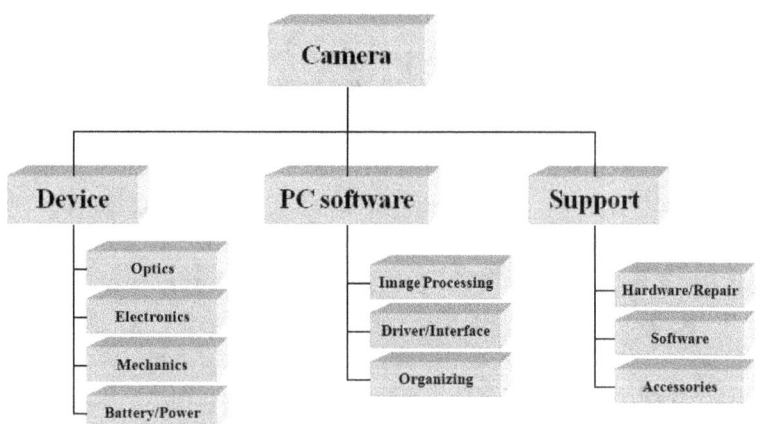

Since the 1980s, systems engineering has gained momentum and now exists as a clearly identified profession in its own right. The primary governing body for the profession is the International Council on System Engineering (INCOSE). Substantial information can be found on the website for this multi-national organisation.

What Do We Mean by a System?

Measuring What You Want to Manage

The point about creating a picture of your system is to be able to identify what it is that you want your system to deliver, and when it will be delivered. There will be an objective, or in system terms, an output. It helps to be able to measure the output so that you can know positively when the system success criteria have been delivered. However, you will also want to know how you are progressing along the route to the goal. Steps or stages in the system development can be seen more easily with the system picture, so that achievement of key parts of the system development can be allocated to time when these elements are in place.

Let's say that you want to achieve faster journey times between London and Bristol, on the Great Western main line railway, in England (GWR). Your objective might be to convert the line to run with electric trains rather than diesel, because they can accelerate and brake quicker, and have the incidental benefits of improved sustainability and lower cost. That is a complex system involving trains, infrastructure, and power supply, amongst many other things. It is important to establish how you will measure where you are in your system development so that you can manage the way forward: how many electric trains are needed in total; how many have been built so far; what percentage of the infrastructure on the route has been fitted with catenary wires to provide the power to the trains; what is the power station capacity in the west of England, to provide the additional power requirement; and how many megawatts does it need to increase by to run the trains; and so on.

The GWR marketing people will want to know how much the journey times are improved by, how many more trains per hour will be provided, and how many extra seats will be available as a result of the improvements. Ticket prices will increase, but look at the benefit that the infrastructure improvements will provide.

Thus, measurement is important in two respects: firstly, to help manage progress towards your goal, and secondly, to be able to identify when you have reached it.

Chapter 2 helps to get a picture of your system drawn up. It will enable you to create your system model.

**There is a website for this book:
www.sensationalsystemsbook.com**

This provides access to a summary of the tips given in this book for successful system delivery. It will also give access to a workbook to help you develop the techniques on which this book is based.

The author, Geoff Miles, is a Chartered Engineer, and runs a coaching, mentoring, and consultancy business. He is also available for teaching and speaking engagements.

**Details can be found at:
www.geoffmilesconsulting.com**

Chapter 2

Defining the System

"Just as the system of the sun, planets and comets are put in motion by the forces of gravity, and its parts persist in their motions, so the smaller systems of bodies also seem to be set in motion by other forces, and their particles to be variously moved in relation to each other and, especially, by the electric force."
– Isaac Newton

In this chapter, you will learn how to define your system in terms of inputs, outputs, boundaries, external influences, and the environment that your system will work in.

System Boundaries

Astronomers are used to working with systems defined only by the limits of their imagination. System engineers generally need to be more constrained in their vision of the systems that they will work with. Consider the example of the systems structure for the different elements associated with an aircraft below.

Sensational Systems

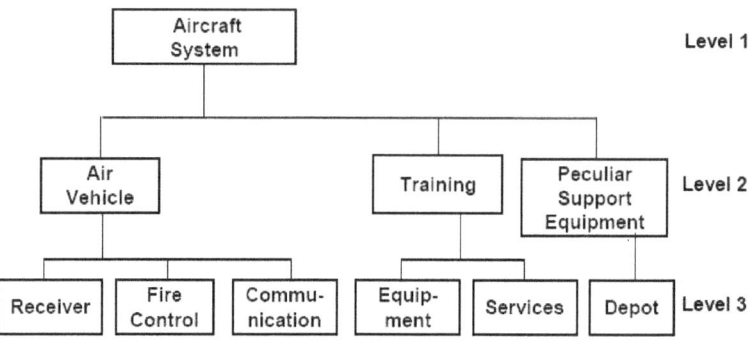

This is a two-dimensional diagram of a system that a military commander might need to think about and manage in order to secure a military objective using air power. The boundary of this system is wide ranging and, for the purposes of the mission commander, it is acceptable to look at the system on three levels. However, when we look at the contribution made by the fire control engineer, we can see that his interest will be at level 3, and the boundary for the fire control system will be at its interface with the other systems contained within the air vehicle.

In Chapter 1, we established the importance of understanding the scale and complexity of your system, and how we need to get a clear picture of this in the form of a system model.

Let's say that our objective is to have a boiled egg for breakfast. Our system model might look like this:

Defining the System

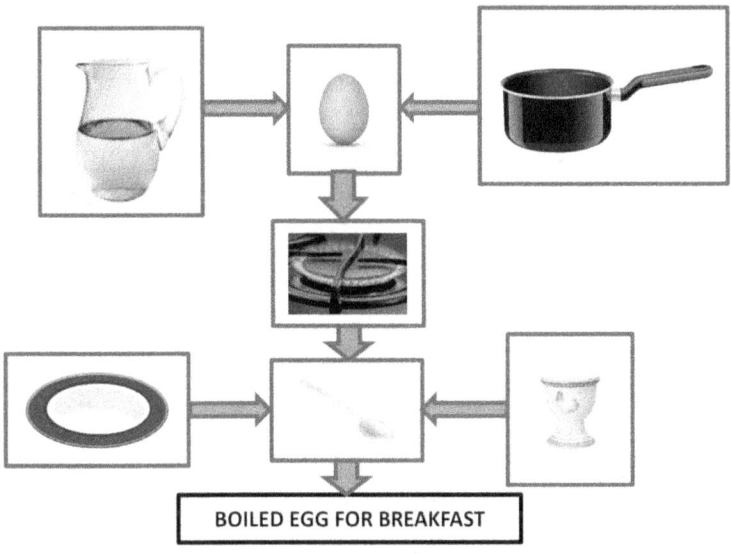

For this system to work, we've had to make a number of assumptions to establish the boundaries for the model. For example, we need to heat water, and we've assumed that we have a hob supplied with gas, and the ability to light it. The model helps us visualise the system and to let colleagues know what it is we are talking about. More importantly, it stimulates thinking about the boundaries and assumptions we are making.

The following sections use the boiled-egg-for-breakfast (BEFB) model to illustrate the points being made.

Defining Your Inputs

As can be seen from the system model, the BEFB system has a number of inputs: an egg, water, saucepan, heat source, plate, egg-cup, and spoon. This simple example is helpful in illustrating the importance of establishing the system boundaries. We are not concerned about where the egg comes from or how the saucepan is made. However, the saucepan should be of sufficient size to contain the egg, and to be able to cover it with water. This would all be written down as a set of input requirements, which would include the diameter of the plate to hold the egg-cup, and a suitable heat-resistant spoon, capable of lifting the egg out of the boiling water and into the egg cup at the end of the cooking time.

And what about the toast? In this example, we have decided that the production of a slice of toast, cut into soldiers, to dip into the egg when it is being eaten, will be part of a different system. A more complex system, called *breakfast production,* might include toast and coffee, but the model helps us to understand the boundaries of the particular system that we are concerned with.

We can then produce our system requirements, which will define the inputs required.

Defining Your Outputs

Our BEFB system has a single output; namely, a boiled egg. However, even in this simple example, the output needs a better definition to be of practical use. I like my boiled egg to be soft, but you might like yours to be hard. The cooking time is the obvious variable, but our current system model does not include a means of determining how long the egg has been cooking for. So, perhaps the model should include

Defining the System

an egg-timer. Cooking time for a soft-boiled egg is also influenced by the size of the egg and its temperature at the time of starting to cook. Is it a large one, straight out of the refrigerator, or a small one, already at room temperature?

It is evident that a system output requirements specification will be needed if we are to expect an acceptable and consistent output from our system.

Knowing the precise output expected from a system, which is part of a more complex system, is important so that problems with the output of the larger system are avoided. In the earlier example of the rail electrification requirement, we know that the nominal voltage available at the catenary contact wire will be 25kV (25000 volts). This will vary downwards within a range, depending on distance from the feeding substation. It may also vary upwards, if regeneration is available from trains braking on adjacent tracks. This may give a voltage range of between 20kV and 28kV for the traction power electrical system output.

Trains that will use the electrical power system output will need to be designed to deal with this voltage variation as an input to its propulsion system. It will still need to achieve its own output requirement of being able to run at 125 miles per hour (one of the train's output requirements), regardless of the input voltage.

As I said, early in Chapter 1, you will need to establish success criteria for your system. A key part of that is defining the objectives and, more specifically, defining the outputs that will deliver that success.

Identifying External Influences

You probably know from personal experience that individuals can be influenced by people around them. Most systems have similar tendencies in that they can be influenced by other systems around them. For the BEFB system, think about the heat source for the water: if it is gas, is it natural gas, propane, or butane? Each has a different calorific value and, therefore, will take different times to heat a given quantity of water. Where will you be cooking the egg? Water boils at a lower temperature at higher altitude, so are you having breakfast as you trek up onto Mount Everest?

Electronic systems can be influenced by adjacent magnetic fields, which will cause interference to outputs, unless attention is given at the design stage to protect the system circuits with, for example, filters or shrouding.

You can see from these two scenarios that it is important for you to examine in some detail the surroundings in which you expect your system to work, in order to deliver the objectives that you have set for it, both reliably and fault free.

Environmental Effects

You will know that the general environment in which we work will influence our performance and well-being. Sitting

Defining the System

on a drawing pin will be decidedly uncomfortable to start with, and we would probably change seats or remove the pin before sitting down again. But if we couldn't do either of those things, conceivably, we could remain sitting on the pin, and we would tolerate the pain; it's not life-threatening, just really uncomfortable. We could then do what we sat down to do. The chances are that what we delivered would not be as good as it might have been if we had been sitting on a well upholstered seat, free of spikes.

Hard systems are like people in this respect. Give an electronic system a hot and damp environment to work in, and it probably would not be as efficient, durable, or reliable at delivering its required output, as one that was given a climate-controlled enclosure to work in.

Thinking through the impacts of vibration, dust, air quality, and similar environmental threats, will be important at the design stages for your system. For soft systems, cyber threats, especially those related to hacking and data theft, will be vital to eliminate at best, or mitigate as far as possible. These aspects of production of successful systems are dealt with in more detail, in Chapter 4, "What could possibly go wrong?"

**There is a website for this book:
www.sensationalsystemsbook.com**

This provides access to a summary of the tips given in this book for successful system delivery. It will also give access to a workbook to help you develop the techniques on which this book is based.

The author, Geoff Miles, is a Chartered Engineer, and runs a coaching, mentoring, and consultancy business. He is also available for teaching and speaking engagements.

**Details can be found at:
www.geoffmilesconsulting.com**

Chapter 3

System Stakeholders

"Consultations, impact assessments, audits, reviews, stakeholder management, securing professional buy-in, complying with EU procurement rules, assessing sector feedback – this is not how we became one of the most powerful, prosperous nations on earth. It's not how you get things done. So I am determined to change this."
– David Cameron, UK Prime Minister, 2010–2016

This chapter is about creating and managing a comprehensive list of system stakeholders.

Who Wants Your System?

You know what you want your system to deliver, but who wants it? Maybe it is you who wants a boiled egg for breakfast, or maybe your partner. How about an electrified railway line? It is so important to establish, at the outset, who wants it: i.e. who are your system stakeholders?

From the inception of systems engineering as a discipline in America, in the 1950s, and its subsequent evolution into a global professional discipline, involvement of stakeholders has been shown to be vital. Because the many facets of our lives revolve around systems of one form or another, systems engineering touches them all.

Chapter 8 covers system assurance and the notion of the V life-cycle for managing your system development and production stages. The life-cycle model discussed there is divided into 8 stages:

1. Concept
2. Feasibility
3. Option selection
4. Outline design
5. Detailed design
6. Building and construction
7. Testing and commissioning
8. Acceptance and handover

Stakeholders can be involved in any or all of these stages, as well as in the system use, maintenance, operation, and subsequent decommissioning of the system. As I mentioned in Chapter 1, systems can be hard or soft, simple or complex, and deal with virtually anything we do in life, so identifying stakeholders is fundamental to achieving success with systems.

Increasing rail line capacity by improving line speed is maybe a government commitment enshrined in the franchise contract for the Great Western Railway Company—or perhaps it is the rail infrastructure owner (Network Rail) who wants to sell more train *paths*, or the rail services operator (GWR) who wants to generate more

System Stakeholders

revenue. You need to be clear about the benefits that your system is intended to deliver; and hence, the interest each stakeholder has in your system.

Stakeholders can be divided into two groups: there are internal, which are those involved with the system development and delivery; and external, which are those affected by the system, who may or may not derive any benefit from it. We need to understand who these individuals and groups are, and what their motivation might be in your system.

Initial identification of who your system stakeholders are, from the outset, whether they are internal or external ones, is very important. Stakeholders can help in setting objectives and success criteria, and where appropriate, contributing to content of the engineering requirements for your system. Identification of the objectives, outputs, and benefits that you will deliver for your stakeholders, helps to establish what the inputs might be. In the BEFB example, the input is an egg, which the requirement might specify as a large chicken's egg at room temperature. The stakeholder is your partner, and the required output is soft-boiled. Job done!

Users, Owners, and Maintainers

So, who will use your system? It may be obvious for an electrified railway, but what about a remotely controlled stair climbing robot? That might not be so obvious. Bomb disposal squads would welcome something that avoids their people coming into direct contact with unexploded devices. And people, who have impaired mobility and cannot easily climb stairs unaided, would benefit. How about piano movers? It could be important to have a range of systems all working on the same principle but having a range of sizes or different scales. So, make the list of intended users.

It is worth exploring what is meant by users: is your system one that needs an operator, such as a machine of some sort? A computer-controlled, metal turning lathe to manufacture precision valve bodies is an example of a complex system that requires an operator to deliver its output. A snack vending machine is also a complex system, but to produce output needs only a user (who might input some money or simply a good kicking to achieve a result!). The stakeholder for the machine tool is likely to be the precision engineering factory owner who requires long life and reliable output from the investment. On the other hand, one of the key requirements of the snack vending machine is likely to be vandal resistance, coupled with low maintenance costs.

The users may not be the owners; and for systems that are sold through retailers as consumer durable products (e.g., electric toasters), it is important to understand who the stakeholders are for these. Key requirements for toasters are likely to be attractive design, together with electrical safety. These requirements will be driven by the brand marketing team, whose main interests may be the reputation of the company, coupled with maximising profit. In these examples, the requirements are not mutually exclusive, and to some extent, can be incorporated into system design, at little or no additional cost. The important thing is to identify the stakeholder requirements at the concept design stage, so that redesign, rework, or retrofitting costs can be avoided prior to the system being delivered.

Then there is system maintenance to be considered. Maintenance is looked at in more detail, in Chapter 9, but here we are considering maintenance from the stakeholder's point of view. Cars are an ideal case to examine: fifty years ago, car owners were expected to check oil, water, and tyre pressure and condition on a daily basis; and then change the oil at 2 or 3 thousand mile intervals, and brakes and clutch at 20 thousand miles. Nowadays, it is taken for granted that cars will go for thousands of miles without lifting the bonnet, or even taking a cursory glance at the tyres. As a minimum, servicing takes place at 12 thousand miles, or when the car's computer decides that some of the consumables have reached the end of their design life.

Competitive pressures on car manufacturers have focussed attention on what the user stakeholders (customers) want, and what will distinguish them from the competition in a crowded market. Maintenance costs are a key part of this, both in terms of the replacement part and the time (and hence, cost) of carrying out the replacement. The use of plastic panels, able to absorb low speed impact without damage, and metal panels that bolt on rather than being welded in place, are a couple of examples where cost is reduced, both for the manufacturer and the owner. And the knock-on benefit to the user's cost of insurance cannot be underestimated. If damaged cars are quick and cheap to repair, then lower insurance premiums can reflect this.

Soft systems are many and varied. Most readers of this book will probably own a smart phone and be familiar with the notifications of software upgrades as a phenomenon of their devices. Who are the stakeholders for these so-called upgrades? The phone user is likely to benefit from the bug fixes, but why were the bugs there in the first place? Possible causes are lack of system development and testing prior to release, the need to be first to market, or just carelessness (or couldn't care less) by the manufacturer. Then there are the upgrades to introduce more features into the software—sometimes it is baffling as to who benefits from these. How many users already do not use more than 10% of the features already provided? Agencies responsible for safety have a big stakeholder input to systems where human life could be threatened. It has been said that there is more computing power in a single Dreamliner first class seat, than in the first moon landing rocket, and that a modern jet aircraft has at least 100 identified faults in its various systems every time it flies. Someone has made the decision to pass the aircraft as fit for service, despite its many problems. Systems critical to flight safety will be the ones that are scrutinised in detail, but those affecting the ability of a customer to view an in-flight movie may not receive the same level of scrutiny. The next chapter considers risks and the use of standards set by regulatory stakeholders to mitigate or eliminate them, in more detail.

System Stakeholders

Who Will Benefit From It and Who Will Pay For It?

Beneficiaries of systems have been mentioned in earlier examples of systems, and from the stakeholder's perspective, they can be a single person or one of the many designers, producers, owners, operators, or users. However, in the modern world, it is convenient to review benefits from two aspects: capitalism and socialism. Someone has to pay for any system, either in terms of money or in terms of time and intellect.

An inventor is a person who thinks of something that does not currently exist, but is able to create it from a collection of existing things, or produces a concept based on simple physics and chemistry. The internal combustion engine, largely as we know it today, is an example of this. Surprisingly, the first engine system designs were around in the 17th century, but it wasn't until 1876 that a group of inventors was granted the patent for the compressed gas, four-stroke internal combustion engine. Obtaining a system to deliver motion from exploding gases was born!

It is arguable but true that we have all benefited from the internal combustion engine system. Gottlieb Daimler (one of the joint holders of the first engine patent) benefitted from the sale of his intellectual property to designers and manufacturers of motor cars, bought by the car's users—capitalism at its essence.

To understand how socialism provides the driver for system development, it may help to consider the development of the endoscope. Endoscopy is defined as the examination of the inside of the human body by using a lighted, flexible instrument called an endoscope. In general, an endoscope is introduced into the body through a natural opening, such as the mouth or anus. Although endoscopy can include examination of other organs, the most common endoscopic procedures evaluate the oesophagus, stomach, and portions of the intestine.

The self-illuminated endoscope was developed at Glasgow Royal Infirmary (one of the first hospitals to have mains electricity), in Scotland, in 1894, by Dr John McIntyre, as part of his specialisation in investigation of the larynx. A major step forward came in the 1950s when fibre-optic light sources were discovered and developed. Most recently, with the development of even smaller miniature cameras and data transmission networks, endoscopy has found new applications, both inside and outside the human body.

Development of medical systems, which help in sustaining life and prolonging active life, generally can be classed as social systems. Stakeholders tend to be governments that fund academic research, particularly those that relate to diseases such as cancer and mental issues in humans. The subsequent effects of such system developments can lead to commercial applications, the income from which can be used for further research and development.

Knowing who will benefit from your system will help you understand not only who will pay for it but also how much they might be prepared to pay. This will help you determine what your system should look like. Characteristics, such as

System Stakeholders

size, quality, and maintainability, will be influenced by the money available from the funder.

Convincing Lord Sugar, and Dragons in their dens, to invest some cash to derive the benefits from your system, should be an exciting prospect for you, and an essential one if you are funding it from your own resources.

Internal and External Stakeholders

In any system, there is likely to be more than one stakeholder. Even with boiling an egg for breakfast, you are the internal stakeholder, and your partner, the intended recipient of your output, is the external stakeholder. How about the involvement of the chicken that laid the egg, the water company that provided the liquid to boil the egg in, and the power company to heat the water.

When planning your system, you need to understand who can directly affect it as a stakeholder, and who can indirectly affect it. You need a comprehensive list that you can categorise into direct and indirect internal stakeholders.

In the same way as internal stakeholders, there are external ones to consider. Does your partner prefer their egg hard or soft? What voltage does the railway electrification scheme need to operate at?

Environmental impacts must be considered by law for many systems, and this brings many external stakeholders into publicly funded works, like road and rail systems. The scale and complexity of managing external stakeholders should not be underestimated, both in terms of cost and time.

Consider the system for importing electric power from France to the UK, using an undersea cable. Simply put, the model of the proposed system might look like the following:

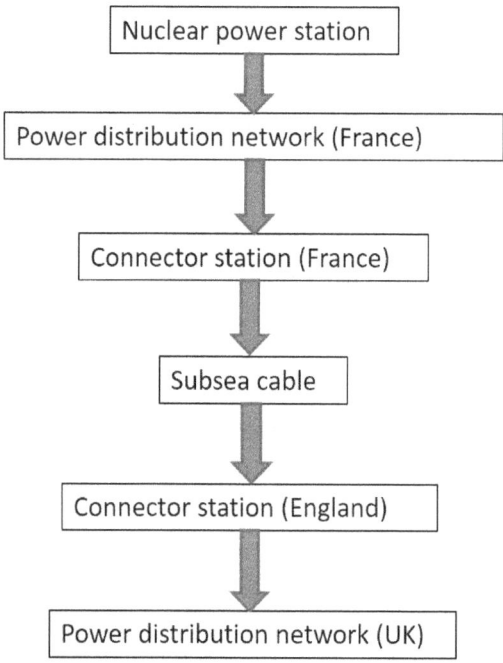

Such a system would involve internal stakeholders from the company that will own the inter connector stations and subsea cable, direct stakeholders from the power distribution networks on both sides of the English Channel, indirect external stakeholders in the two national regulatory authorities, directly impacted stakeholders whose land would be occupied by the connector stations, and indirect external stakeholders who would be impacted by the environmental effects of construction and operation of the high voltage power equipment.

In the UK, planning and environment legislation provides guidance on how developers need to go about stakeholder management for their systems. Formal records need to be kept, and where compulsory purchase of land is involved, or where major developments of such facilities as new and upgraded railways are involved, specific Acts of Parliament give legal powers to ensure enforcement of Undertakings and Assurances.

At its simplest, it is important to make a list of all stakeholders for your system, and keep it up to date.

Stakeholder Aspirations

With lists of both internal and external direct and indirect stakeholders available, it is then necessary to consider their aspirations in relation to your system. This analysis will reveal those who will benefit, and those who will be impacted, and whether these impacts will be either positive or negative. I suppose it is human nature that focus is generally on those who are adversely impacted: this will range from those who just dislike change, to those who will lose their homes.

It is important to know what your system will do for which people. Funders will worry about return on investment, and whether increased revenue from shorter journey times, for instance, will meet forecasts. Government agencies may worry about job creation and standards of living. What are your stakeholder's aspirations?

Consultation and communications strategies and plans can be built around this knowledge. These can be linked to risk assessment and mitigation plans and strategies. Where new systems require acquisition of land, particularly where environmental impact assessments (EIAs) predict adverse impacts on your neighbours, legislation provides rules that must be complied with, to ensure that developers meet at least minimum standards for their systems. Management of huge complex systems are beyond the scope of this book, but similar principles apply to boiling an egg for breakfast. For completeness, the following two paragraphs are included to give a flavour of dealing with large infrastructure schemes.

For railways and tramways in the United Kingdom, the UK government introduced the Transport and Works Act. Orders under the Transport and Works Act 1992 (the TWA) can authorise guided transport schemes, and certain other types of infrastructure projects, in England and Wales. Promoters of schemes of this kind often need a range of powers to put their scheme into practice. Under the TWA, a promoter can apply to the Secretary of State (or to the Welsh Government, for schemes entirely in Wales) for an order giving those powers. The order, if made, is known as a TWA order. The powers that can be given in a TWA order can be very wide-ranging. For example, the promoter of a new railway or tramway scheme may need compulsory powers to buy land or to close streets. A TWA order can grant these powers. Putting the scheme into practice could affect people's enjoyment of their property, and affect the environment. Because of this, applications for TWA orders have to follow a set procedure, which allows people to give their views on the proposals.

For schemes of national significance, such as the high-speed rail network development in the UK, powers are granted through specific Acts of Parliament. Stakeholders can voice their concerns at public inquiries, and can obtain undertakings and assurances from the scheme developers, which get enshrined in law. For example, maximum limits can be set for exposure to noise and vibration for properties within the zone of influence of a new high-speed rail line. It would be a criminal offence for the owner of the line to exceed the defined limits.

For the vast majority of systems, stakeholder aspirations will be concerned with cost, reliability, security, safety, quality, maintainability, and other such system characteristics. Identification of these at the system concept stage will help to ensure delivery of a successful system.

There is a website for this book:
www.sensationalsystemsbook.com

This provides access to a summary of the tips given in this book for successful system delivery. It will also give access to a workbook to help you develop the techniques on which this book is based.

The author, Geoff Miles, is a Chartered Engineer, and runs a coaching, mentoring, and consultancy business. He is also available for teaching and speaking engagements.

**Details can be found at:
www.geoffmilesconsulting.com**

Chapter 4

What Could Possibly Go Wrong?

"The real test is not whether you avoid this failure, because you won't. It's whether you let it harden or shame you into inaction, or whether you learn from it; whether you choose to persevere."
– Barack Obama, US President, 2009 to 2017

This chapter is about risks that may threaten your system, and how to manage them successfully.

Consider the Risks

There are those people—let's call them risk professionals—who make their careers out of worrying about what could possibly go wrong. For many people, risk assessment happens subconsciously; it's what keeps most of us alive. We all have built-in levels of risk tolerance, which varies from person to person. Risk tolerance varies for each of us as we go through life. We all know of young *dare-devils*, but why do racing drivers generally retire before the age of 35? And why do some 80-year-olds go sky-diving for the first time? Perhaps the age-versus-risk tolerance graph is bathtub-shaped.

Sensational Systems

It is important to carry out risk assessments for your system, to know what could possibly go wrong and what we might be able to do, either to eliminate the risk or at least know how it might be mitigated or reduced to its *reasonably practicable* level.

Risks can also be considered in terms of what is at risk in your system. Is it safety, reliability, availability, maintainability, or reputation, for example? What is affected, and how important is that to your system? How much are you prepared to compromise in design or functionality, and how much might you be prepared to pay to eliminate or at least mitigate that risk?

Identify the system risks and categorise them.

Complex systems warrant production of a Risk Management Plan (RMP). There are two important benefits of such a plan. Firstly, to help the system developer throughout the design, production, and testing and commissioning stages of system development to have a clear strategy for reviewing risks, allocating responsibility for eliminating and mitigating those risks, and to check that risks to or caused by the system, to the user, are at their lowest practicable levels.

Secondly, the RMP and the resulting documents, such as hazard analyses and designers risk assessments, will provide evidence to stakeholders, of competent system development. This will be particularly important to owners and funders to demonstrate that sufficient attention has been given to system safety, to avoid downstream litigation for injury and damage caused by the system.

Likelihood of Risks Occurring and the Impact if They Did

Imagine walking down a street and thinking about the risks of getting injured or killed whilst walking. Slips, trips, and falls are the obvious candidates for injury. The likelihood of tripping is quite high, increasing as we get older and more infirm. How about the chance of a piece of aircraft falling on you from 35,000 feet? Quite remote, you might think. And then, somewhere in between, a mugger taking a fancy to your smartphone, or a car mounting the pavement and knocking you down.

Similarly, you need to assess the likelihood of a risk event happening to your system.

Undeterred, you carry on walking down the street. You are aware of the risks and of the likelihood of the event happening, but what would be the implication if it did occur? You might take a tumble if you did slip or trip, and you might fall on your knees or hands and get grazed. Probably the worst that could result is you would fracture or break a bone or two.

You could eliminate the risk by not going walking, or by reporting the uneven pavement to the local authority for repair or replacement. Or you might mitigate the risk by wearing sensible walking shoes, walking with a stick, or using a support frame.

There's not much you can do about a lump of aircraft falling on you: you'd probably be killed instantly. However, with an infinitesimal chance of it happening, you decide you are prepared to take the risk, and carry on walking without constantly peering skywards (and reducing the risk of tripping by looking where you are going!).

Of course, the current phenomenon of smart phones makes the opposite true. In any cosmopolitan city, many people are roaming the streets with heads bent, looking at the small screens of their devices. They have disregarded the risk of an aircraft engine part falling on them and, mostly, have disregarded the risk of bumping in to other pedestrians. The assumption is that the other head-down addict will somehow avoid them. Perhaps we are evolving the sense inbred into bats, which seem to be able to fly in the dark without injury.

Getting mugged could be mitigated by staying in populated areas, not going down dark alleys, not carrying valuable objects (unlikely), making sure wallets and purses are not easily accessible, and so on.

In the same way, risk impact should be assessed for your system. Eliminate the risk of a system component breaking, by ensuring that it is designed to withstand the applied load, and has, for instance, guards for protection against impacts. Reduce the risk of power failure, by having uninterruptable power supplies or back-up batteries and generators.

So, you need to assess the impact of a risk event

happening to your system if it did occur.

Risk Ranking and Tolerance to Them

Risks can be prioritised in a quantified way by creating factors for the chance of a risk happening (likelihood), and for what would happen if the risk did occur (impact or severity), then adding the two factors together. This process will give you a metric to enable you to carry out a risk ranking. This process is detailed below.

The first thing to do is to decide the factoring score for likelihood. A simple grading often used is:
1 – Never
2 – Very unlikely
3 – Unlikely
4 – Likely
5 – Always

Or to put a timescale on the likelihood of a risk occurring, often used for engineering systems:
1 – Once in a hundred years
2 – Once every ten years
3 – Once per year
4 – Once per month
5 – Once per week

Then decide the factoring score for impact. A simple grading often used is:
1 – None
2 – Insignificant
3 – Minor

4 – Major
5 – Disastrous

For systems that are in the public domain, and where injury to people may occur if a risk manifests itself, impact is often classified in relation to extent of injuries caused.
1 – Non-reportable injury
2 – Minor reportable injury
3 – Major injury
4 – Single fatality
5 – Multiple fatalities

Each of the factoring scales (risk ratings) can be varied to suit the system that is being risk assessed. Each risk rating is then built into a risk matrix, populated with the factors added or multiplied together.

The following table gives an example of a simple risk rating matrix:

		Likelihood rating				
		1	2	3	4	5
Impact rating	1	2	3	4	5	6
	2	3	4	5	6	7
	3	4	5	6	7	8
	4	5	6	7	8	9
	5	6	7	8	9	10

Key

	2-4 low risk	Negligible
	5-7 medium risk	Tolerable
	8-10 high risk	Intolerable

Risk ranking grades can be varied (between 5 and ten is normal), and the ratings can be multiplied together, instead of adding, to give greater emphasis on high risk events, and to better distinguish those risks that are intolerable.

Risk Management Process

Your Risk Management Plan (which was discussed earlier in this chapter) should have a simple process for setting up a risk register for the production life of your system. This will be initiated by a hazard identification workshop attended by representatives of the system's direct stakeholders.

The system model normally forms the basis of the review, where each part of the model, including inputs, internal elements, outputs, and the external environment, are examined for hazards and risks.

The Royal Society for the Prevention of Accidents provides a useful five-step process for carrying out a risk assessment:

Step 1: Identify the hazards. In order to identify hazards, you need to understand the difference between a *hazard* and a *risk*.
Step 2: Decide who might be harmed, and how.
Step 3: Evaluate the risks, and decide on control measures.
Step 4: Record your findings.
Step 5: Review your assessment, and update as and when necessary.

These process principles can be applied across a range of systems, but it is important to get down to specifics for the system being reviewed and assessed.

The risk assessment will also vary with the stage of the system development. The V life-cycle model, discussed in Chapter 3, gave the following 8 stages:

1. Concept
2. Feasibility
3. Option selection
4. Outline design
5. Detailed design
6. Building and construction
7. Testing and commissioning
8. Acceptance and handover

An initial risk assessment can be carried out at the concept stage, and then it should be reviewed and updated at least through the outline design, detailed design, and commissioning stages.

For any larger systems, at handover of your system to its owner and users, a risk register will form part of the Health and Safety file, so that operators and maintainers will know of any identified hazards and risks that they will be responsible for managing throughout the system's life.

Hazard and risk management is not a one-size-fits-all activity. Just think of the diversity contained within systems associated with plumbing, fire safety, police searches, working in confined spaces, and construction work. It is

What Could Possibly Go Wrong?

important to tailor your risk registers to the systems being developed.

Hazard identification checklists can help stimulate discussion in hazard workshops. General headings might include:

- Manual tasks
- Storage
- Lighting
- Hazardous chemicals
- Compressed gas
- Machinery and plant
- Electrical
- Noise, vibration
- Ventilation and heat
- Confined spaces
- Ladders, excavations, access, and walkways
- Housekeeping, amenities, and office environment
- Signage, first aid, fire detection and suppression, and emergency plans
- Security, including cyber security
- Personal protective equipment

Why do a risk assessment? The answer is that a risk assessment will protect your workers and your business, as well as comply with law. As for when to do a risk assessment, it should simply be conducted before you or any other employees carry out work that presents a risk of injury or ill-health.

If you are a business owner, then ideally, a person from your organisation needs to attend risk assessment training. This will ensure that this person is competent within your organisation, and will gain abilities such as hazard identification, and the ability to categorise and evaluate risk(s). These abilities will allow a suitable and sufficient risk assessment to be conducted within your own organisation.

Five steps to risk assessment can be followed to ensure that your risk assessment is carried out correctly. More details of the five-step process are as follows.

Step 1: Identify the hazards

In order to identify hazards, you need to understand the difference between a *hazard* and a *risk*. A hazard is *something with the potential to cause harm*, and a risk is *the likelihood of that potential harm being realised*.

Hazards can be identified by using a number of different techniques, such as reviewing your system design, walking round the workplace, or asking your employees.

Step 2: Decide who might be harmed, and how

Once you have identified a number of hazards, you need to understand who might be harmed, and how.

Step 3: Evaluate the risks, and decide on control measures

After identifying the hazards and deciding who might be harmed, and how, you are then required to protect the people from harm. The hazards can then either be removed completely, ideally through the system design process, or the risks controlled so that the injury is unlikely.

Step 4: Record your findings

Your findings should be written down (in the UK, it's a legal requirement where there are five or more employees), and by recording the findings, it shows that you have identified the hazards, decided who could be harmed, and how, and also shows how you plan to eliminate the risks and hazards.

Step 5: Review your assessment, and update as and when necessary

You should never forget that few systems (or workplaces where systems are used) stay the same, and as a result, your risk assessment should be reviewed and updated when required.

The UK Health and Safety Executive definition of a risk assessment is:

"...a careful examination of what could cause harm to people, so that you can weigh up whether you have taken enough precautions or should do more to prevent harm..."

Risk Mitigation and Ownership

In the previous section, we saw how to rank any risks identified within your system, both in terms of likelihood of the risk occurring, and the severity of impact if it did occur. Taking these two factors together, we then came to a view on whether or not the risk is tolerable.

Ideally, from the outset, all the risks in your system are tolerable, but if they are not, then something must be done to make sure that they become so. The first thing to do is to examine whether the risk can be eliminated by design (or redesign!). But if your system includes something that gets hot, and the risk identified is that the system user may inadvertently come into contact with the hot system element, then what mitigation might be put in place to prevent this happening? Obviously, in the hot case, some heat resistant shielding could be installed, or a guard put in place. Warning signs may help; at worst, training the system user in the likely dangers that person may encounter, should help.

Whatever mitigation is possible, it should be recorded on your hazard log: what needs to be done, and at what stage of the system development process will the action be taken. Someone also needs to take ownership of the risk, and be given responsibility for carrying out the mitigation action. The aim is to ensure that all system risks are tolerable before the system is put into use. Identifying the risk owner is vital where a risk moves through the eight-stage system lifecycle; this is because the person responsible for minimising the risk may change. Take, for example, a system that will boil an egg automatically. The designer comes up with the notion that an egg-shaped basket is attached to a conveyor, which moves through a bath of boiling water, dwelling for a definable time before moving the egg to a collection point. The designer's hazard log records identified risks, including the fact that the system will have hot water and a hot egg output. The water will be heated by an electrically powered immersion element, so risk of electrocution will be also identified.

The initial risk responsibility will be allocated to the designer, who will ensure that the design includes a decorative, insulated, and heat-resistant casing around the water container. It will also include an interlock for the conveyor, so that the lid must be closed to enable the conveyor to move, thereby avoiding the risk of the operator getting fingers trapped. Where the mitigation is incorporated into the design, the responsibility for risk management will pass to the test and commissioning engineer, who will ensure that the design meets the expected outcome. A risk review might result in a recommendation to fit warning labels (risk of scalding; danger, isolate from mains before removing electrical cover; etc.). Then, when the system passes to the user, there will be an operations and maintenance manual, which will include a safety warning, and the risk responsibility will move to the operator to comply with the safety instructions.

In larger, more complex systems, particularly those involving construction, the Construction (Design and Management) Regulations (CDM) provide a clear structure for how risks can be minimised and dealt with. Clients involved with these systems have legal duties of compliance. The next section gives more detail of this.

Compliance With Requirements

In the last fifty years, European Union legislation has been produced as a series of regulations, which many systems are obliged to comply with, and to demonstrate compliance by testing, especially where the public may come into contact with the system. Goods coming into Europe from outside the EU must be able to show that they comply equally with those produced inside the EU.

In the UK, the Health and Safety at Work etc. Act 1974 (also referred to as HSWA, the HSW Act, the 1974 Act or HASAWA) is the primary piece of legislation covering occupational health and safety in Great Britain. The Health and Safety Executive, with local authorities (and other enforcing authorities), is responsible for enforcing this Act, and a number of other Acts and Statutory Instruments relevant to the working environment.

The Act lays down general principles for the management of health and safety at work, enabling the creation of specific requirements through regulations enacted as Statutory Instruments or through a code of practise. For example, the Control of Substances Hazardous to Health Regulations 2002 (COSHH), the Management of Health and Safety at Work Regulations 1999, the Personal Protective Equipment Regulations 1992, and the Health and Safety (First Aid) Regulations 1981, are all Statutory Instruments that lay down detailed requirements. It was also the intention of the Act to rationalise the existing complex and confused system of existing legislation.

Since the accession of the UK to the European Union (EU) in 1972, much health and safety regulation has needed to comply with the law of the European Union, and Statutory Instruments under the Act have been enacted in order to implement EU directives. In particular, the Act is the principle means of complying with Health and Safety Framework Directive 89/391/EEC on health and safety at work. Further important changes to duties in respect of articles and substances used at work were made by the Consumer Protection Act 1987, in order to implement the Product Liability Directive 85/374/EEC.

If your system is being designed, produced, and used outside of the umbrella of European legislation, then other major countries will have legislation intended to protect workers, users, customers, and other stakeholders to your systems. It is essential that you become familiar with what these are, and how your system meets (and ideally fully complies with) those requirements and legislative obligations.

Throughout the world, standards set out requirements, which many systems should meet or exceed. In the UK, British Standards have been in place for many years, but since the accession of the UK into the European Union, they have gradually been replaced by Euro Norms (EN) standards. Many other industrialised countries set national standards. For example, as the voice of the US standards and conformity assessment system, the American National Standards Institute (ANSI) empowers its members and constituents to strengthen the US marketplace position in the global economy, while helping to assure the health and safety of consumers and the protection of the environment. The Institute oversees the creation, promulgation, and use of thousands of norms and guidelines that directly impact businesses in nearly every sector: from acoustical devices to construction equipment, from dairy and livestock production to energy distribution, and many more. ANSI is also actively engaged in accreditation, assessing the competence of organizations determining conformance to standards.

Similarly, Japanese Industrial Standards (JIS) specifies the standards used for industrial activities in Japan. The standardization process is coordinated by the Japanese Industrial Standards Committee (JISC), and published through the Japanese Standards Association (JSA). The Japanese Industrial Standards Committee is composed of many nationwide committees, and plays a vital role in standardizing activities in Japan.

If your system is to be used as part of an aircraft system, or one of the stand-alone systems within an aircraft, it may be helpful to know that some specific industries use standards as requirements. The National Aerospace Standards (NAS) are voluntary standards developed by the aerospace industry. Subject matter experts from Aerospace Industries Association (AIA) member companies participate in committees and working groups to develop and maintain the NAS library, which currently contains over 1400 active

What Could Possibly Go Wrong?

standards. These standards cover a wide variety of subject areas including:

- NAS parts (bolts, rivets, washers, screws, nut plates, pins, knobs, etc.)
- Safety Management Systems (NAS9927)
- Non-destructive Test Personnel certification (NAS410)
- Hazardous materials management (NAS411)
- Foreign Object Debris (FOD) prevention (NAS412)
- Cutting tools (drills, reamers, end mills)
- Airport Operations (NAS3306)
- Trade Compliance Standards (TCS)

In addition, the International Air Transport Association (IATA) is the trade association for the world's airlines, representing some 280 airlines, or 83% of total air traffic. They support many areas of aviation activity, and help formulate industry policy on critical aviation issues.

IATA's mission is to represent, lead, and serve the airline industry.

Representing the airline industry, they improve understanding of the air transport industry among decision makers, and increase awareness of the benefits that aviation brings to national and global economies. Advocating for the interests of airlines across the globe, they challenge unreasonable rules and charges, hold regulators and governments to account, and strive for sensible regulation.

Leading the airline industry for over 70 years, they have developed global commercial standards, upon which the air transport industry is built. Their aim is to assist airlines by simplifying processes and increasing passenger convenience, while reducing costs and improving efficiency.

Serving the airline industry, they help airlines to operate safely, securely, efficiently, and economically, under clearly defined rules. Professional support is provided to all industry stakeholders, with a wide range of products and expert services.

Clearly then, if your system is associated with the aerospace industry, there is plenty you can learn and be supported in with getting compliance for your system. And there are many other specific market-related trade bodies and agencies that can similarly help with system success. Finally, in this chapter, it is worth getting an understanding of the term, *reasonably practicable*, because this phrase appears in documents, regulations, and guides associated with risk and risk management.

In the UK, what is reasonably practicable is a question of fact. The Court of Appeal held in 1949 that: *"in every case, it is the risk that has to be weighed against the measures necessary to eliminate the risk. The greater the risk, no doubt, the less will be the weight to be given to the factor of cost."* – Lord Justice Tucker

And:
"Reasonably practicable is a narrower term than 'physically possible,' and seems to me to imply that a computation

must be made by the owner in which the quantum of risk is placed on one scale and the sacrifice involved in the measures necessary for averting the risk (whether in money, time, or trouble) is placed in the other, and that, if it be shown that there is a gross disproportion between them—the risk being insignificant in relation to the sacrifice—the defendants discharge the onus on them." – Lord Justice Asquith

These principles are a requirement of the Management Regulations, and apply to all industries, including construction. They provide a framework to identify and implement measures to control risks on a construction project. The general principles of prevention are to: (a) avoid risks; (b) evaluate the risks which cannot be avoided; (c) combat the risks at source; (d) adapt the work to the individual, especially regarding the design of workplaces, the choice of work equipment, and the choice of working and production methods, with a view, in particular, to alleviating monotonous work, work at a predetermined work rate, and to reducing their effect on health; (e) adapt to technical progress; (f) replace the dangerous by the non-dangerous or the less dangerous; (g) develop a coherent overall prevention policy that covers technology, organisation of work, working conditions, social relationships, and the influence of factors relating to the working environment; (h) give collective protective measures priority over individual protective measures; and (i) give appropriate instructions to employees.

System risks is a huge subject, with many facets. To have a successful system, it is necessary to apply some effort to managing risks along the lines of the principles set out in this chapter. For simple systems, this will be enough, but for more complex systems, some specialist help may be needed, and a more thorough understanding of legal obligations may be important. The next chapter should help in determining what will be needed.

**There is a website for this book:
www.sensationalsystemsbook.com**

This provides access to a summary of the tips given in this book for successful system delivery. It will also give access to a workbook to help you develop the techniques on which this book is based.

The author, Geoff Miles, is a Chartered Engineer, and runs a coaching, mentoring, and consultancy business. He is also available for teaching and speaking engagements.

**Details can be found at:
www.geoffmilesconsulting.com**

Chapter 5

Systems Within Systems

"Being busy does not always mean real work. The object of all work is production or accomplishment, and to either of these ends there must be forethought, system, planning, intelligence, and honest purpose, as well as perspiration. Seeming to do is not doing."
– Thomas A. Edison

This chapter is about understanding dependencies outside your system, and interdependencies within systems. Having understood those, it is then possible to explain to others how things fit together.

Simple or Complex

At the start of this book, we looked at system types and whether the one that you are involved in is simple or complex. There are no hard rules to determine which type falls into which category; it largely depends on the system application you are considering.

Using the example of a car, it could be part of a transport system, which includes public as well as private transport elements, moving people into and out of town centres, which is the subject of the system scrutiny. Or it might be the means of commuting to and from your place of work, and it is that outcome that you need to gain assurance of. Or it could be that you need to understand automatic transmission systems of the type fitted to your car, and what might go wrong with that system.

The actual car that features in each of these systems can be simple or complex, depending on where you draw the system boundary. Hence, the importance of determining which system you are concerned with. In Chapter 2, we looked at the need to define your system boundary so that you are able to know what inputs are required and where the expected output will arrive. If you are a transport planner, and the system you are interested in is moving people in and out of the town centre, then you can consider not only private cars but also buses, taxis, trains, motorcycles, and pedal bikes, in that they are all simply a means of moving people. Inputs might be the number of people moved per unit and the speed at which they travel. The car, in this example, is a simple system (a means of moving people). But if you are a gearbox specialist, you probably are not interested in the media systems contained within the transport mode, and your system boundary will focus on the power input from the engine to the gearbox, and the power output from the gearbox to the driving wheels. In this case, the gearbox is the simple system with the boundary just around the gearbox.

Systems Within Systems

Interdependencies

There are interdependencies between each element within your system, and between systems within complex systems. There is no need to get distracted by the internal workings of each system if all you need to achieve is a defined outcome from your system.

Using the car example, you know that the car engine provides power, which needs to be delivered to the wheels in contact with the road. The link between the two is the transmission system and the drive train. A simple model can be drawn of the four parts of this complex system:

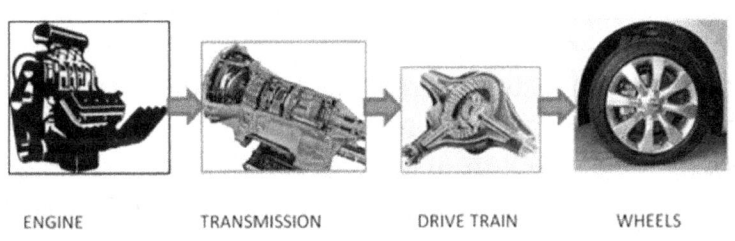

ENGINE   TRANSMISSION   DRIVE TRAIN   WHEELS

Unless we are concerned about the inner workings of the automatic transmission, it is enough to understand that there needs to be a way of connecting the front of the gearbox to the back of the engine, and that the maximum power output from the engine can be taken by the gearbox consistently without problems. The same is true for the transmission connection to the drive train, and the drive train connection to the wheels. Further interdependencies can be deduced from knowledge of the amount of suspension movement of the wheels, which has to be accommodated by the connection between the drive train and the wheels.

By reducing complex systems to manageable elements with simple inputs and outputs, system engineering and assurance can be more efficiently handled.

This is an important lesson to learn, especially when your stakeholders may not all be familiar with the internal workings of an automatic gearbox (or similar complex systems). The ability to make complex elements appear simple to the lay person is a valuable attribute.

If you are a salesperson for a domestic water purifying system, you may understand the chemical, electrical, biological, and thermodynamic processes necessary to obtain good tasting pure water from the sea or from brackish well sources, using reverse osmosis principles. But your clients, who are mostly everyday householders, do not and probably would not understand the detail, even if you gave them as much detail as you have.

However, if you used system engineering principles of creating a system model with defined boundaries and clear inputs and outputs, you could maybe sell more of your product by making what is a very complex process appear to be simple.

Eliminating Complexity and Interdependence

See what you can make of the system model for a domestic water purification system, which follows below:

Systems Within Systems

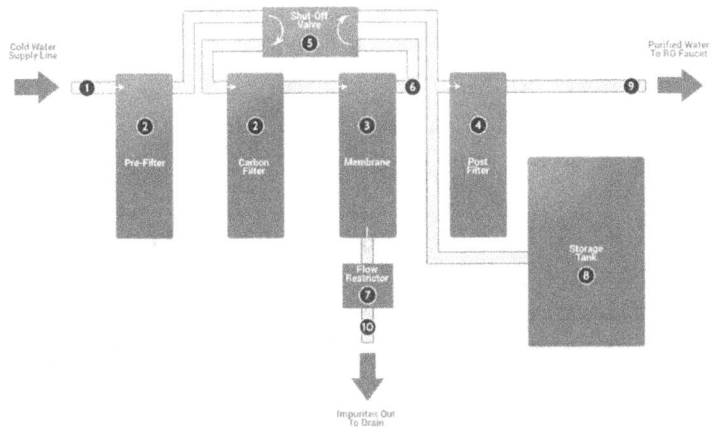

Basic components of a Reverse Osmosis (RO) system:

Cold water line valve: A valve that fits onto the cold water supply line. The valve has a tube that attaches to the inlet side of the RO pre-filter. This is the water source for the RO system.

Pre-filter(s): Water from the cold-water supply line enters the RO pre-filter first. There may be more than one pre-filter used in a RO system, the most common being sediment and carbon filters. These pre-filters are used to protect the RO membranes by removing sand, silt, dirt, and other sediment that could clog the system. Additionally, carbon filters may be used to remove chlorine, which also can damage the RO membranes.

Reverse Osmosis membrane: The RO membrane is the heart of the system. The semi-permeable RO membrane is designed to remove a wide variety of both aesthetic and health-related contaminants. After passing through the membrane, the water goes into a pressurized storage tank where treated water is stored.

Post filter(s): After the water leaves the RO storage tank, but before going to the RO faucet (clean drinking water tap), the treated water goes through a final *post filter*. The post filter is usually a carbon filter. Any remaining tastes or odours are removed from the output water by the post filtration *polishing* filter.

Automatic shut off valve: To conserve water, the RO system has an automatic shut off valve. When the storage tank is full, the automatic shut off valve closes to stop any more water from entering the membrane, and blocks flow to the drain. Once water is drawn from the RO faucet, the pressure in the tank drops; the shut off valve then opens to send the drinking water through the membrane, while the contaminated wastewater is diverted down the drain.

Check valve: A check valve is located in the outlet end of the RO membrane housing. The check valve prevents the backward flow of treated water from the RO storage tank. A backward flow could rupture the RO membrane.

Flow restrictor: Water flowing through the RO membrane is regulated by a flow restrictor. There are many different styles of flow controls, but their common purpose is to maintain the flow rate required to obtain the highest quality drinking water (based on the gallon capacity of the membrane). The flow restrictor also helps maintain pressure on the inlet side of the membrane. Without the additional pressure from the flow control, very little drinking water would be produced because all the incoming water would take the path of least resistance and simply flow down the drain line. The flow control is most often located in the RO drain line tubing.

Storage tank: The standard RO storage tank holds from 2–4 gallons of water. A bladder inside the tank keeps water pressurized in the tank when it is full. The typical under-counter RO tank is 30cm diameter and 40cm tall.

Faucet: The RO unit uses its own faucet, which is usually installed on the kitchen sink. Some areas have plumbing regulations requiring an air gap faucet, but non-air gap models are more common.

Drain line: This line runs from the outlet end of the RO membrane housing to the drain. The drain line is used to dispose of the wastewater containing the impurities and contaminants that have been filtered out by the RO membrane.

Hopefully, with the help of a simple system model and brief explanatory notes, you are able to understand the basic concepts of how a reverse osmosis water purification system works. The point is that whatever system you are setting out to make successful, reducing or eliminating complexity should be one of your important goals.

Interdependence can be defined as the dependence of two or more people or things on each other. If you study biology, you'll discover that there is a great deal of interdependence between plants and animals. *Inter* means *between*, so *interdependence* is *dependence between things*. You may often see *interdependence* used to describe complex systems. Marriage creates a state of interdependence between spouses. If your dog provides you with love and happiness, and you provide your dog with food and walks (and love and happiness), then your relationship with your dog is one of interdependence.

In systems, interdependencies are a source of risk, particularly the risk of failure, because if one part of the system is dependent on another part of the same system, failure of one will result in failure of the other, and so you may double the risk of overall system failure. Eliminating interdependencies within your system should be a design goal if you want to achieve a reliable system. Of course, this should not be confused with system dependencies where there is an activity flow through your system. A sequential series of activities, which takes the system input and processes it through a set of activities to produce the desired output, is part and parcel of most physical systems. Thinking about the interdependence between plants and animals helps to appreciate how interdependencies in some systems, like those established in ecology, are vital for survival. Plants depend on animals for carbon dioxide, which they need to achieve photosynthesis; bees and butterflies help in pollination and dispersal of seeds. Earthworms and other organisms make the soil fertile for plant growth. Animals depend on plants for food, medicines, supply of oxygen, and removal of carbon dioxide. Shelter-plants are homes for a variety of animals. Humans rely on plants and animals to provide material for buildings, furniture, and clothing. Both animals and plants depend on sun, air, water, and soil for their growth and sustenance. Thus, all organisms depend on one another for their survival.

Establishing Interfaces and Responsibilities

From the earlier car systems model, we can see that some interdependencies are similar between different parts of the model. For example, the amount of power transmitted to the wheels is related to the amount of power produced by the engine. Of course, there are some losses (due to friction, heat, and noise), but they are predictable and can be taken into account when determining the anticipated output of the system.

It may be important to the claimed features of your system to compensate for interdependencies and build them in at the design stage.

Effective managers are usually people who can *see the wood for the trees*, who can make elaborate things appear simple, and can draw clarity from complexity. To become good at system engineering and integration, you need to develop these skills. And analysing interdependencies is a good way to start.

If something is internal within an element of your system, with little or no external influences on it, it can perhaps be treated as a system on its own (For example, do you really need to know how an automatic transmission system works, as long as you know what inputs are needed to get the required outputs?), and so, in your system model, it can be simply shown as a block.

If you are looking at current flow within your railway electrification system, the train that uses the electrical power can be shown as a single block, which draws current from the overhead conductor wire and sends it back to the substation, using the running rail. The interdependency in such a system could be related to the energy needed to achieve travel between Bristol and London within 90 minutes, rather than whether the train's electric motors are linear or asynchronous. We can see that it is worthwhile to spend some time with your system model to simplify it, focusing on the overall system objectives to identify what is important.

Inevitably, in most systems, you will end up with some interfaces, and someone will need to be allocated a responsibility for dealing with it. A literal nuts and bolts example will help to understand the principles of this. In the railway electrification system, it is likely that the power supply to the overhead wires (known as the catenary system) will be connected to a power transformer contained within a substation. The transformer will be protected by a circuit breaker so that an electrical fault in the catenary system will not affect the transformer.

The electrical distribution company (EDC) will supply the substation with its equipment, and the catenary system will

Systems Within Systems

be supplied by the catenary installation contractor (CIC). One of the interfaces in the electrification system has been identified as the cable connection between the catenary system and the circuit breaker in the substation. Straightforward, you might think; the EDC supplies the circuit breaker inside the substation, and the CIC supplies the cable to connect to it. But who supplies the nuts, bolts, and washers to join the two items together? The details of this (and of all the other similar interfaces) should be included in a systems interface specification. This will provide information, including the size of holes and centreline spacing in the circuit breaker connection busbar for the bolts, the cable termination lug details, the cable diameter, and its minimum bend radius, and the route that the cable will take through the substation wall to reach the connection location. Interfacing parties will be identified, and responsibilities allocated to the party best able to carry out the connection (the CIC in the case of the nuts, washers, and bolts).

The time spent in producing the Systems Interface Specification will be invaluable in ensuring a successful system.

Helping Others to Understand Your System

Unless your system is entirely for you alone, there will be other stakeholders involved, both directly and indirectly. Chapter 3 gives more detail about types and roles of stakeholders. The important thing here is to be aware that you, as the system designer, business manager, or project manager will need to help others understand your system.

It is likely to be helpful if you are able to incorporate the three main topics covered in this chapter into your presentations, leaflets, brochure, explanatory notes, and other such material used to inform and educate stakeholders about your system.

The three main topics covered are:

1. Reducing complexity as much as possible
2. Identifying interdependencies and the relationships between them
3. Establishing the system interfaces and allocating responsibilities for them.

A few words about presentation material and techniques might be worth refreshing here.

The three T's

1. Tell them what you are going to say
2. Tell them
3. Tell them what you said

The five P's

1. Prepare – identify the desired outcome for the presentation.
2. Purpose – have a clear understanding of the intended purpose of the material.
3. Practise – set aside enough time before any presentation to practise and get the right length of presentation for the time available to give it.

Systems Within Systems

4. Present – Have a warm-up ritual that *gets you in the zone*. Know how to leverage body language to boost your confidence and audience engagement.
5. Post-op – self-critique after each presentation; get external feedback, and feed forward.

There are a good number of other processes that you can search for on Google to help you with presentations and explanatory material, but the theme, which most of them share, is making complex matters appear simple to the majority of people.

I suggest that it is worth remembering that the World Wide Web went live to the world on 6th August, 1991. Sir Tim Berners-Lee invented it in 1989, and it arguably became the catalyst for the explosive growth of the digital revolution, which continues today. Imagine being the system inventor trying to convince the majority of the world that this system was worth parting with good money to invest in such an ethereal notion; especially when the dot-com bubble burst in 2001. A feature of the 1990s was the digital divide with the people of older generations, who thought (and maybe still think) that a mouse is still a rodent, and the web is somewhere for a spider to catch its next meal. The digital divide is getting greater, as the gap widens between rich and poor people, both within developed and developing countries.

Helping others to understand your system is important, even though it may not be as impacting on society as the World Wide Web—at best, eliminate, or at least reduce jargon.

As Albert Einstein once said, *"Everything must be made as simple as possible. But not simpler."*

There is a website for this book:
www.sensationalsystemsbook.com

This provides access to a summary of the tips given in this book for successful system delivery. It will also give access to a workbook to help you develop the techniques on which this book is based.

The author, Geoff Miles, is a Chartered Engineer, and runs a coaching, mentoring, and consultancy business. He is also available for teaching and speaking engagements.

Details can be found at:
www.geoffmilesconsulting.com

Chapter 6

Building Your System

"Another important consequence in the arrival of digital technology and its facilitation of feedback is that we can look at large systems and recognize them once more not only as part of ourselves, but also as components that can change...
Now though, we live in a world where text is fluid, where it responds to our instructions. Writing something down records it but does not make it true or permanent. So why should we put up with a system we don't like simply because it's been written somewhere?"
– Nick Harkaway, British Novelist

This chapter is about developing a clear, logical strategy and plan for bringing your chosen system to life.

System Components and Build Sequence

The V life-cycle we are using throughout this book (and is looked at in more detail in Chapter 8), for system development, has 8 stages:

1. Concept
2. Feasibility
3. Option selection
4. Outline design
5. Detailed design
6. Building and construction
7. Testing and commissioning
8. Acceptance and handover

We've had our light bulb moment to dream up the system concept; we've done our sums and determined that the whole thing is feasible; we've looked at some options for how it might come together, selected one, and drawn up an outline design. We've done a hazard analysis and completed a risk assessment, ironed out a few problems during the detailed design, and now we've got to get on and make the system.

I find it helpful to use a physical type of system in my descriptions and examples, but the principles apply similarly to both virtual systems and those based in software.

During the detailed design stage, we should have generated our system requirements. This is the list of the components that will be used to make our system. It will include all of the items, their specifications, and the quantity needed.

Specification is an important word because it encompasses many things. Generically, there will be two groups of system attributes under the specification umbrella. Firstly,

Building Your System

the system functional requirements document (FRD), which specifies such features as:

- System definition
- System characteristics
- Design and construction
- Safety and security
- Documentation
- Operation and use

Secondly, there should be the system specification document (SSD). This will build on the FRD to provide more detail about the system requirements. Typical headings within the SSD may include:

- System definition
- System description
- System diagrams
- System interfaces
- System characteristics
- Performance characteristics (including those for hardware and software where applicable)
- Physical characteristics
- Reliability
- Maintainability
- Environmental
- Human factors and ergonomics
- Technical assurance and safety assurance
- Quality requirements
- Standards to be complied with

A few words about the relationships between the client, funder, and the system builder may be helpful here. If the client is not intending to be the user of the system, and their objective is simply to make some profit from filling a gap in the market for a particular system, they may not go beyond the functional requirements document stage. They will agree with the funder the selling price that the market will bear, the profit that each wants to make, together with other cost estimates for elements such as production, testing, commissioning, logistics, packaging, storage, and distribution, to arrive at the maximum cost of bringing the system to market. A high level FRD will be developed, and third-party manufacturers will be asked to bid for the supply using their own SSD. The range of permutations in initial system documentation is immense because of the range of systems that are developed. As a result, you are free to select what works best for your system situation.

Defining the Work Breakdown Structure

Your system model will give you a good basis for getting a work breakdown structure (WBS) in place.

We are at the 6th stage of the project V life-cycle (i.e. building and construction); we've got a detailed design and a system specification, so we have a clear understanding of the system requirements, and we know what needs to be done. The WBS is designed to show you it will come together. A simple example is shown below.

Simple Work Breakdown Structure

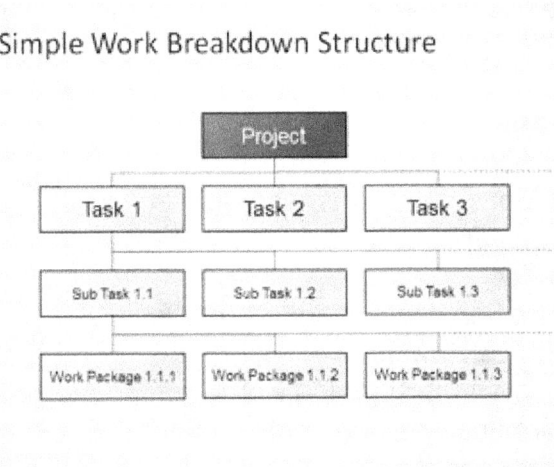

The essence of the WBS is to break the project system into a series of high level tasks, then to divide each task into a number of sub-tasks, and then divide each sub-task into individual work packages. The number of divisions of the project (normally referred to as level 0) into level 1 tasks, level 2 sub-tasks, and level 3 work packages, depends on the complexity of the system under review. The more complex the system, the more levels are needed to make the work packages of appropriate size for manageability.

In larger systems, it may be worth considering the engineering discipline responsible for each of the primary tasks. For example, if your system requires bespoke software, it is probably best to allocate a specific work package to software development.

A typical Software Requirements Document might contain the following headings:

- Purpose
- Applicable documents
- Definitions, acronyms, and glossary
- Problem description and assumptions
- Data flows
- Information content and structures
- Design constraints
- Functional partitioning, programme narratives, and performance requirements
- Testing requirements

This will allow management effort to be directed specifically to ensuring that the system software is available and thoroughly tested prior to being needed to drive your system hardware.

The production of software is very important in all but the simplest of systems, and a number of books focus entirely on effective management of software production. However, the generation of software can be visualised from the V life-cycle adapted for the stages that software development goes through, as shown on the next page.

Building Your System

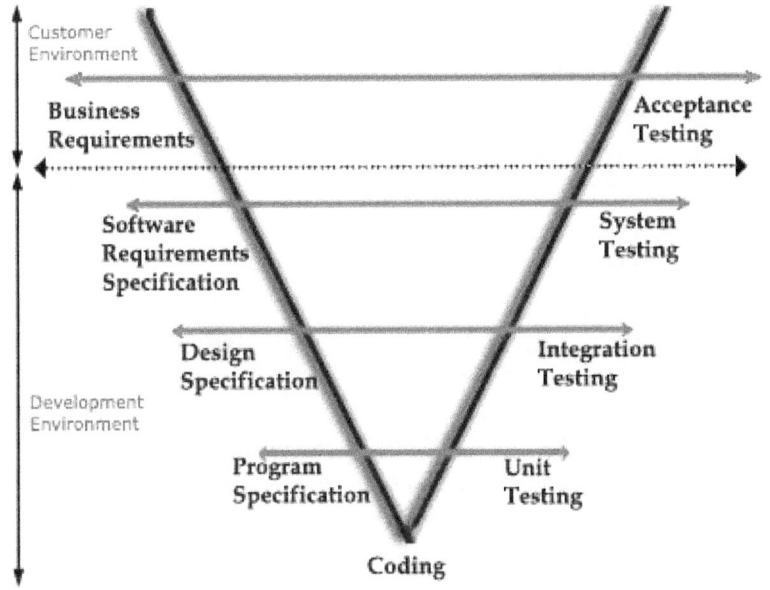

This can follow the WBS pattern adapted for software development typically as follows:

Management
- Project management
- Configuration management
- Verification and validation
- Quality assurance

Programming
- Coding
- Unit testing
- Reviewing

Requirements and design
- User requirements
- Software requirements

- Design
Testing
Computer usage
- Hardware
- Software
- Facilities
- Support

Back in Chapter 2, we had a simple model for a system that would produce a boiled egg for breakfast (the BEFB Project).

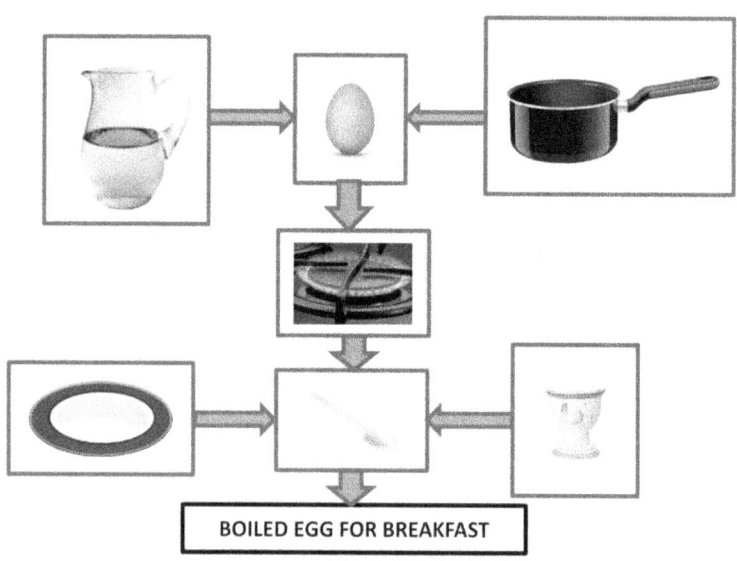

If we use this to produce our Work Breakdown Structure (WBS), it might look like the following three-level structure:

Building Your System

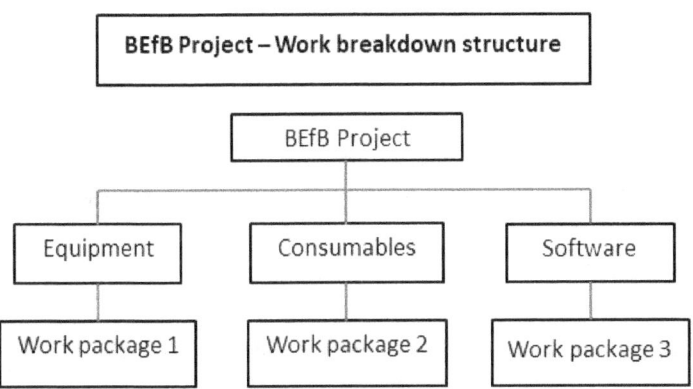

Lead Times and Priorities

Having got our Work Breakdown Structure (WBS) sorted, we now need to think about how long system production will take, and what we need to do first to get it started.

Examination of each of the identified work packages (WP) will give us the tasks that need to be carried out. For our BEFB project, we have three WPs:

Work package 1: This is the equipment work package, and the purpose of this is to assemble the items of equipment required. As part of our system design, we should have prepared a schedule of materials and equipment. For WP1, we might have allocated the following items:

- Water container
- Saucepan
- Gas cooker

- Plate
- Spoon
- Egg cup

For this example, we can assume that all the items of equipment are available from stock in my house, except for the egg cup, which I will need to procure. I will buy one on-line, and the delivery time is three working days.

For WP2, the consumables package, my schedule of materials lists:

- Gas
- Water
- Egg

I already have metered gas and water supplies to the house, but currently, no eggs. My on-line shopping delivery, which includes six eggs, is scheduled to arrive in five days' time.

The software pack is to be obtained through WP3. My social media account has recognised that I had previously ordered eggs, and had presented me with an advert for an App for obtaining the *perfect boiled egg for breakfast*, for a cost of only £3.99, which it will run on my smartphone. To get this, I must allocate some time to log in to the iTunes App store (or Google Play store), pay for the App, and download it. I must then teach myself how to use it (settings will include the size of the egg to be boiled and my preferred hardness of the yolk when cooked). This will take me at least a day!

Building Your System

Ideally, I would employ a Project Manager to manage this system production task, but the BEFB project is a simple system, and as I have some basic project management skills (and this book), I will do that job myself. Software is available to help me, such as Microsoft Project or Primavera, but these work on the principle of presenting information as Gantt charts. A generic example is shown below.

Gantt

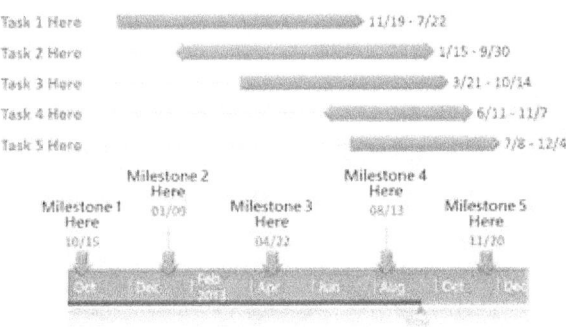

A Gantt chart (more commonly known as a bar chart) gives you a visual representation of the tasks that need to be carried out, together with an estimate of the duration each task will take to complete. It is also able to establish milestones within the development and production schedules. Milestones are normally points in time when significant events are completed or are able to start. More about this is included in Chapter 8.

An important feature is that a bar chart is able to show dependencies between tasks. In the BEFB project, for example, you will not be able to start cooking breakfast until the egg delivery is completed in five days' time.

With the knowledge of task duration and the dependencies between tasks, you are quickly able to construct a clear plan of action. What is known as the *critical path* can be established. This is shown with bold arrows in the diagram below. The critical path is the series of activities whose durations, when added together, will determine the overall time that the system building will take from start to finish. This is absolutely crucial to know, when others are depending on your output.

As a system designer and manufacturer, it will be disingenuous if you promise your client the system will be delivered in 10 days' time, if your critical path shows that production will take 15 days.

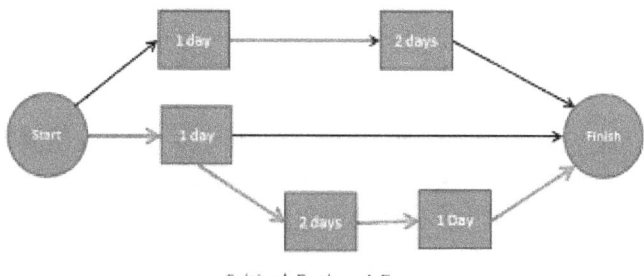

Critical Path = 4 Days

Building Your System

An extract of a critical path (the lighter bars) from a more complex project plan is given below.

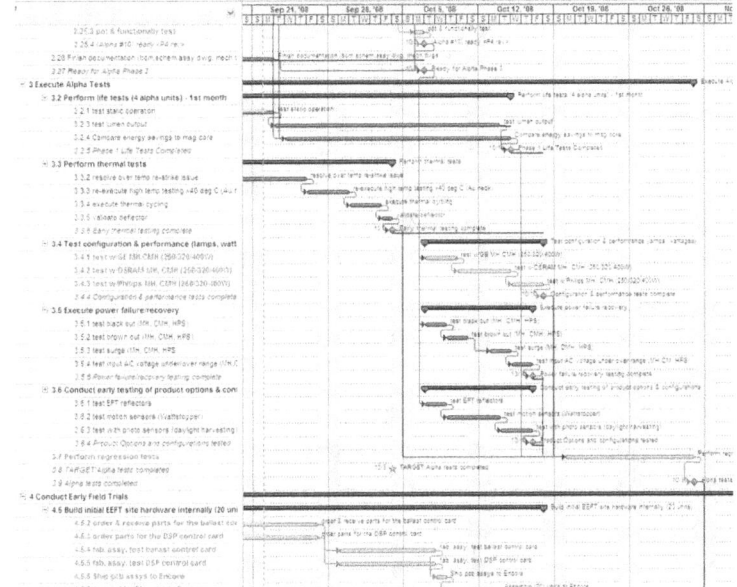

There are a number of different methodologies that you can use for representing lead times and priorities for system development plans, depending on which organisation does the teaching, and what your personal preferences are, but the basic principles are the same.

Resources and Logistics

At some stage in building your system, you will need to direct attention towards managing resources. The main resources are time, money, and people.

It's often said that more than half of new businesses fail during the first year of trading, but the American Small Business Association (SBA) states that only 30% of new businesses fail during the first two years of being open, 50% during the first five years, and 66% during the first 10 years. Whatever statistics might actually be true, the SBA, in an article on small business failure, lists the following reasons for failure, from Michael Ames' book, *Small Business Management*:

- lack of experience
- untrusted sales representative
- insufficient capital
- poor inventory management
- over-investment in fixed assets
- business's finance mismanagement
- poor business location
- poor credit arrangement management
- unexpected growth
- engaging in the wrong business niche
- inability to recover from a major business interruption

It is thought that some 80% of business failures are due to lack of cash. It is really important to focus on money and cash flow in system development. Your work breakdown structure can help you avoid being in the majority, so far as business failure statistics are concerned. It can be adapted for use as a Cost Breakdown Structure, where the cost of the tasks and activities within each work package can be estimated and shown against a time dimension, so that you can see when money needs to be paid out. If you will only be paid at the despatch of your system to the client, then you will be able to see what cash you need to have in order to reach the end point.

Financial planning is paramount for many businesses where products have relatively long lead times before they come to market, and especially when the system products contain expensive materials and sub-systems that have to be procured well in advance of the sales value returning. Cash flow forecasting could be the subject of a book in itself, so I recommend that you seek more detailed advice if money management is not one of your current core skills. Many of the software products that support project management have add-ons, which include money management elements. This allows you to combine time, cost, and labour resource planning for the basis of the one work breakdown structure. This can be really helpful if you need to carry out *what-if* analyses; the effects of a change in one leg of the resources triangle can be seen in the other two legs.

The third resource to manage effectively, if you are going to deliver a sensational system, besides time and money, is people. People (also known as manpower or human resources) are the key to success in virtually all aspects of life. In the systems context, people are essential because of the creative thought and logical reasoning that they bring to all of the activities associated with system design, development, manufacture, testing, commissioning, packaging, distribution, sales, and marketing. And this is where the system work breakdown structure can help again.

Sensational Systems

At the lowest level in your system WBS, you have the activities and tasks that need to be carried out in order to bring your system into being. By listing these out on the vertical scale, you can put the different skill types that people will need to have to build your system, on the horizontal scale of the matrix, and then set about populating the boxes.

A simple example of this is a building environment management system, for which the system model is given below.

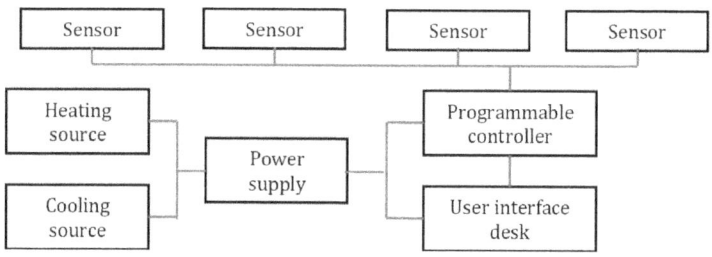

The WBS should result in work packages for sensors, power supplies, control systems, and climate control systems. By analysis of the labour resources needed for each work package, we would be able to produce a labour skills resource plan similar to that shown on the next page.

Building Your System

Work package	Activity	Electrician	Electrical test engineer	Software engineer	HVAC engineer	System integrator
Sensors	Installation	1				
	Testing		1			1
Power supply	Installation	4			1	
	Connection to switchboard	1				
	Testing		2			1
Control system	Produce software code			10		
	Installation			2		1
	Testing		1			5
Climate control system	Installation	1	2		10	1
	Ducting				5	
	Testing		2			3
Days per resource type		7	8	12	16	12

This matrix shows us that for the installation, testing, and commissioning stage of the building environmental management system project, we need five categories of labour resource, for a total of 55 days. If the average day rate paid for this type of work is £400, the total labour cost will be £22,000. This principle can form the basis for resource management for your system.

The final part of this section concerns logistics. The Cambridge Dictionary defines logistics as *"the careful organisation of a complicated military, business, or other activity so that it happens in a successful and effective way."* For sensational systems, logistics strategies and management can be considered in two groups. The first group is the logistics associated with the production of the system, and the next few paragraphs cover that. The second group is operational support logistics for when your system is put to use, and is expected to work correctly every time. That aspect is covered in Chapter 9, the chapter dealing with operation and maintenance.

Let's imagine that we were employed by the Airbus consortium in the 1990s, as production logistics specialists, and we were given the task of developing a logistics strategy for the A380 aircraft production. We were told that finished aircraft will weigh over 280 tonnes empty, and have a maximum take-off weight of 570 tonnes when full of fuel and with 800 passengers on board. Components for the aircraft are to be built in Germany, Spain, United Kingdom, and France. Structural assembly will be carried out in Toulouse, France, and final fit-out and finishing will be done in Hamburg, Germany.

Clearly, it would have been a major logistical challenge for the system to build the A380 aircraft. But the challenge was successfully met, and by mid-2018, more than 220 had been built. Although the production rate has recently slowed down, it reached a peak of one aircraft per week. For interest, the logistics strategy adopted included the construction of a fleet of roll-on-roll-off ships and barges, the construction of port facilities, and the development of new and modified roads to accommodate oversized road convoys. The map on the next page shows the routes taken by the large assembly components of the aircraft.

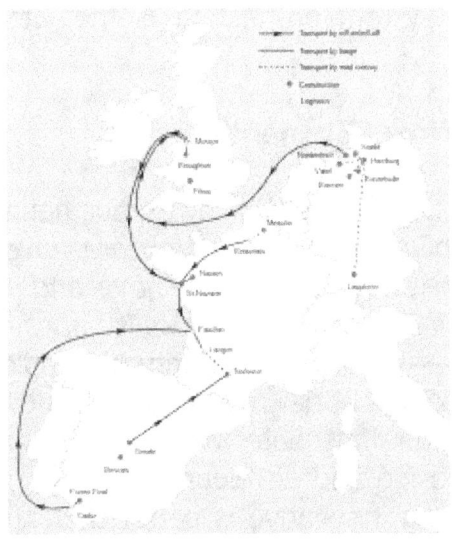

Whilst the logistics strategy for production of your system may not require such massive endeavours to be successful, the principles are the same. Think about where components and assemblies will come from; what special materials handling equipment will be needed; what dimensions and weights will need to be dealt with; where testing will be carried out; and what special features, such as trackwork and power supplies, might be required; and so on.

Relate the logistics plan to your production bar chart so that you are able to show when, for example, a bespoke lifting frame for your carbon filter module might be needed. If the lead-time for the lifting frame is six-weeks, then you could see when the latest order placement date was. Careful, meticulous, detailed planning of activities in advance of system production is the vital element of success.

Proven or Novel Elements

David Burkus, an award winning author, defines innovation as *"The application of ideas that are novel and useful. Creativity, the ability to generate novel and useful ideas, is the seed of innovation, but unless it's applied and scaled, it's still just an idea."* The majority of people are curious, and are thinking that new ways of achieving any particular result is a natural attribute. When this thinking is applied to system design and development, it is normal to think that something innovative is better than the old way of doing things, but we all know of innovation that has failed. The systems used over time for water distribution make an interesting topic for demonstrating the argument about whether to use proven or novel components in your system design. The Romans first piped water through houses and public buildings quite successfully using clay pipes. Later, the Americans, without the clay sources, successfully used timber pipes. In Britain, elm wood was used for wooden water pipes, but these were found to leak and quickly rotted. With the invention of cast iron, pipes using that material were made, but these were both expensive and found to produce evil-looking iron-stained and tainted water, and so were not popular.

Cities, such as Dublin, Manchester, and London were expanding rapidly, and at the turn of the 19th century, city engineers were exploring how to get fresh water piped to its citizens, with a water distribution system. It was discovered that Guiting Stone, found deep in the Cotswold Hills in South West England, uniquely yielded large blocks of stone, which could be bored to make stone pipes.

In 1805, the Stone Pipe Company (SPC) was set up to supply these rapidly enlarging British cities with drinking, and every other sort of water, through pipes of clean, solid, and *pure* stone. When SPC production targets shot up, Guiting Stone was chosen as the company's single source of stone. For a few frantic years, a massive manufacturing enterprise operated, with more than 30 tons of bored pipes leaving the works each day. Expensive plans soon followed to run first a major canal, and then a long tram road, into the SPC works, to further facilitate the creation and transport of even more pipes.

Then, in July 1812, at first, in London, the pipes suddenly failed on a massive and terminal scale. They had simply proved incapable of withstanding the higher, now steam-engine driven, water pressures being demanded. These were needed to provide *high service*, to the tops of the fashionable houses it was hoped could now be supplied in these cities. Many of Britain's most famous engineers had been deeply involved: James Watt Senior (the steam engineer), William Murdoch (the pioneer of gas lighting), and John Rennie (the famous civil engineer and SPC chief engineer and share-holder), with tram road wagons supplied by the famous Butterley Company in Derbyshire. Bankruptcy, including of one major London bank, then followed, in one of the first cases of *systems failure* in British engineering history.

Interestingly, a tour of some of the villages in the vicinity of the old Stone Pipe Company works, in Gloucestershire, will reward you with the sight of garden walls with circular holes in them; these having been constructed from surplus materials output salvaged from the factory when it was forced to close.

Clay and improved quality cast iron continued to be used for many years for water pipe systems. More recently, however, most modern pipe systems are made in plastic materials, as this is a readily available material.

There are several lessons for systems engineers to learn from this example. Firstly, proven components have less risk of failure than novel ones. Secondly, novel components of a system need to be thoroughly tested to the limits within which the system is designed to operate (in the case of stone pipes, maximum internal pressure that the steam pumping engine can deliver, although this may not have been foreseen when the system was first designed).

However, there are two important drivers for the use of novel components in systems: firstly, when your particular niche in the systems market is crowded, and competing on price is not possible, you need some features that set your system apart, and innovation and novelty may be the way to gain market share. Careful risk assessment is required to balance the risk of failure against the opportunity presented by getting a bigger slice of the market.

Secondly, as I said at the start of this section, human curiosity is natural, and it is the way we have evolved through time. The explosive growth in smart phone technology gives an indication of how a novel system can change life for the better. It is interesting that human nature, in that case, develops slower than the technology, and there is perhaps a growing trend for mobile phone handsets to revert to simpler systems to fulfil their original basic function of allowing people to communicate easily with each other.

In the next chapter, we will be looking at system testing and commissioning. To prepare for that, it might be helpful to review some typical system engineering terms, and what their true meaning is.

System engineering term	Actual meaning
Close project coordination	We know who to blame
Major technological breakthrough	It sometimes works OK, but it looks very hi-tech
Preliminary operational tests were inconclusive	It blew up when we switched it on
Test results were extremely gratifying	We were so surprised when the stupid thing worked
All new	Parts not interchangeable with the previous version
Energy saving	Achieved when the power is switched off
Low maintenance	Impossible to fix if broken

**There is a website for this book:
www.sensationalsystemsbook.com**

This provides access to a summary of the tips given in this book for successful system delivery. It will also give access to a workbook to help you develop the techniques on which this book is based.

The author, Geoff Miles, is a Chartered Engineer, and runs a coaching, mentoring, and consultancy business. He is also available for teaching and speaking engagements.

**Details can be found at:
www.geoffmilesconsulting.com**

Chapter 7

Testing and Commissioning

"The business of life is to endeavour to find out what you don't know from what you do; that's what I called 'guessing what was on the other side of the hill.'"
– The Duke of Wellington

This chapter is about getting your system testing and commissioning organised to deliver a sensational system.

Measurements

The importance of testing your system should not be under-estimated, and it is best that this is completed as soon as possible. The V life-cycle model for systems has three main stages:

Stage 1 – concept and outline design
Stage 2 – detailed design and construction
Stage 3 – testing, commissioning, and acceptance

In terms of the total cost of bringing your system to market, expenditure on system development can typically be 10% of total cost of bringing the system to market spent to get to the end of stage 1; 80% for getting stage 2 done, and 10% for completion of stage 3. It follows that the earlier you can find and rectify a problem, then the least it will cost to fix overall.

It may be considered contradictory for stage 3 to be called testing, commissioning, and acceptance, when I have suggested that you should test as early as possible, but you need to think about the purpose of the testing and what you are trying to measure as part of the tests. The concept design will provide a set of objectives or goals that you want your system to deliver. It may not be sensible to spend 90% of the cost of development before you gain confidence that the objectives will be met. Chapter 4 looked at the risks associated with your system, and an obvious one is that the system fails to deliver what you wanted. To assess this, you need to quantify the desired output and have some means of measuring your expectations.

Earlier in this book, I used the example of a system for reducing the rail journey time between major cities. The system proposed was rail infrastructure electrification and procurement of a fleet of electric trains. Journey time reduction would be achieved in part by increasing the train acceleration and braking rates, and introducing the ability to run safely at higher speed than existing trains. Note that rail travel is a complex system involving many features and subsystem elements, such as reducing the number of intermediate stops between city centres, improving signalling systems, and track infrastructure alignment improvements. This example deals only with the train speed element.

The benchmark for the existing journey time can be measured: existing (diesel powered) rolling stock accelerates at 0.2 metres per second (m/s), can reach a top speed of 125 miles per hour (mph), and will brake at up to 0.5 m/s. The system for journey time reduction will have the objective of improving acceleration to 0.4m/s, increasing maximum speed to 140mph, and braking rate to 0.7m/s.

The train system designers will show that they will achieve these rates by improving five features of the train: the number of powered axles on the train (the coefficient of friction governing the amount of force that can be transmitted from the train wheels to the rail track could be the subject of a whole book by itself, but suffice it to say here that 0.3 can be assumed); the weight of the train reduced from 500 tonnes to 320 tonnes; the number and size of brake discs; the aerodynamic efficiency of the train shape; and the ability of the train to tilt on curved track.

The train system project manager faces a dilemma with the Train Operating Company (TOC) because the commercial contract requires the journey time reduction objective to be demonstrated before any money is paid to the train manufacturer. They know that it typically takes thirty months to build a new train, and each new train will cost on average £15 million. The train manufacturing company cash flow will not be able to finance the cost of any rework if the objectives cannot be met first time.

Their solution is to agree a test strategy with the TOC, using a combination of computer simulation and modelling, benchmarking with equivalent systems and subsystem testing, using rigs and prototypes to demonstrate the ability of their company to meet the objective within six months of contract award, well in advance of major expenditure on production items.

The primary underlying lesson to be learnt from this example is that determining the metrics for quantification of your system objectives is fundamental to successful system delivery.

Test Plans and Schedules

Test plans and schedules are needed for effective development of successful systems to validate and verify that the system meets its objectives. The Project Management Body of Knowledge defines validation and verification as follows:

Validation: The assurance that a product, service, or system meets the needs of the customer and other identified stakeholders. It often involves acceptance and suitability with external customers.

Verification: The evaluation of whether or not a product, service, or system complies with a regulation, requirement, specification, or imposed condition. It is often an internal process.

Testing and Commissioning

Generally, test plans should be drawn up to deliver two primary purposes: one inward looking and the other outward facing.

The overall test plan for a system will cover the production life-cycle, and will identify the points of testing at each stage of the process. The "what" of the test plan can be linked to the "when," by relating it to the bar chart for production. The "how" will be given in a test procedure drawn up specifically for each test point.

Test procedures will vary to suit the particular system or sub-system being tested, the type of industry, and any in-house company requirements, but generically, a test procedure might contain the following section headings:

- Title, document number, originator, approver, and revision history
- Introduction and context
- Submission date requirements, including a copy of the procedure used, the test data recorded, and the test report
- Abbreviations used and glossary
- System data
- Test safety requirements
- Test description
- Pass criteria
- Instrumentation
- Test conditions and calculations
- Test sequence
- Test results
- Observations and comments

- Signature page for tester in charge and witnesses

Testing plans, test schedules, test procedures, and test reports are all important documents in the process of systems assurance. More details of this are given in Chapter 8.

Planning for carrying out system testing is important from the aspect of ensuring that sufficient resources of the right type are available when the testing is scheduled to be undertaken. Resources will include the space to carry out the test; this could be a work bench, a room, a building, an area in the open air, a length of railway line, and so on. Resources will also include the necessary people to carry out the test. This may be a person with special skills, such as an acoustics engineer, or someone with a special license, such as a gas appliance tester. Resources also include test equipment and materials; a load test on a support structure may need large weights to simulate the intended load. For systems involving pressure vessels, suitable pressure testing instrumentation will be needed.

In a railway signalling system upgrade, where the system is part of an operational rail network, conducting the final system test will involve stages of wheel-free testing, so the rail network needs to be closed for the duration of the tests. You may have experienced the *bus-replacement-service* at weekends and bank holidays, due to *engineering works*. This activity needs to be planned months in advance so that train operators can minimise the inconvenience to passengers. In addition, the Tester-in-Charge of a railway signalling system (i.e. the one person who certifies and authorises the signalling system to safely protect a crowded train from collision) is a very scarce resource in the UK. Allocating this vital resource for the correct amount of time, and having a back-up plan in case of ill-health or other non-availability, is an important task for the systems person responsible for delivering a successful rail signalling system.

For your system success, I suggest you develop and manage a comprehensive testing strategy, plan, procedures, and schedules. This will enable you to confidently verify that your system meets its objectives.

Stakeholder Involvement

The outward facing part of assurance is validation (i.e. ensuring that your system achieves the requirements). Normally, the person or organisation that is paying for the system to be designed, developed, built, and commissioned is very interested in seeing that they are getting what they wanted. Chapter 3 considered the range of stakeholders that may have an interest in your system. Here, we need to look at stakeholder involvement in testing and commissioning.

The starting point is production of a test plan. As noted earlier, the test plan will look at the system production life-cycle and identify when specific tests will be carried out. The test plan is then circulated to the various stakeholders for review and comment, and importantly, to identify where they will be required to attend specific tests for sign-off and acceptance. In larger systems, it is normal for the system production contract to have a series of stage payments, agreed at clearly identified milestones, in the production schedule. It is convenient for some milestones to be linked to specific tests, and for stakeholders to be invited to witness the test being successfully carried out so that both parties can sign off the test report as evidence for the payment to be made.

Indirect stakeholders, such as residents who live near to a site where a system is to be installed, may need to be involved as part of an undertakings and assurance commitment made at the time of planning consent being given for the system installation. An environmental impact assessment (EIA) may have been carried out as part of the planning process, where an assurance was given that a power transformer system would not increase ambient noise by more than 3 dBA in the nearest dwellings to the site. Failure to achieve this commitment by the contractor would result in a liability to install triple glazed windows in the affected buildings. In such a case, stakeholder involvement could be a source of major cost for the contractor if the testing and reporting is not carried out thoroughly by experts.

For complex systems comprising sub-systems manufactured by sub-contractors, testing can be divided into a number of different elements. In the high-speed train example, the traction power element is a significant system by itself but is a sub-system part of the total train. Stakeholders associated with the traction power system will need to be involved at different stages of testing. At an early stage of design development, a static simulation test rig may be built, incorporating all of the component parts of the power system. It will then be exercised and evaluated to ensure that the overall system can provide the train with the acceleration and braking capability demanded by the system improvement criteria. There will be a Factory Acceptance Test (FAT) for the first off the production line, and then, when the system is installed in the train, there will be static and dynamic performance testing carried out. As an example, power, signalling, and electro-magnetic compatibility (EMC) engineers will be involved in specific tests to ensure that the thyristors in the power system do not generate harmonic frequencies that may affect the signalling system (which depend on single frequencies for safety protection). Stakeholders from the infrastructure owners, adjacent telecommunications networks, and other railways will also have an interest, and may be involved in final system acceptance.

Simulating Failure and Recovery

In Chapter 4, we considered what could possibly go wrong with your system. By conducting a hazard workshop and analysis, we arrived at a risk register that needs to be managed through the life of the system. As part of the risk mitigation for the reduction, and where possible, the elimination of risks we may have identified, the fact that we did not know exactly what might happen if a failure of a particular type occurred, and what the impact might be if it did, should not be a problem. During our risk management activity, we might have decided to design and conduct specific tests to provide the data required for risk assessment.

The London Underground tube railway network is one of the safest in the world, and this has not come about by accident. Take fire safety, for example. It is the result of meticulous attention to detail, ensuring compliance with fire safety standards, reduction or elimination of materials that support combustion and sources of ignition, plus incorporation of fire detection and suppression systems where there is a risk of fire.

A Tube train seat is an example where testing has enabled the seat design to be optimised for fire safety, whilst still providing a good degree of passenger comfort and resistance to vandalism. Fire is hazardous mainly due to surface spread of flame, which causes blocked fire escape routes, heat to build up in confined spaces, and toxic fumes to be given off by the combustible materials, which causes breathing difficulties, loss of vision, and disorientation. The British Standard 6853:1999 had specific tests to assess a particular seat design for its ability to allow fire to spread (the crib test), and for smoke and fume output (the 3-metre cube test). All seats used in trains designed to work underground had to meet the specified criteria as a minimum. French and German fire safety standards were different, but with the Technical Specifications for Interoperability (TSIs) being gradually introduced throughout Europe after Britain joined the European Union, the British standard was replaced by a Euro norm (BS EN 45545:2013), to harmonise safety requirements.

Railway operating systems provide a good example where testing is used to simulate system failure and recovery. Consider a converging track junction, where two trains are able to approach a single track from different directions.

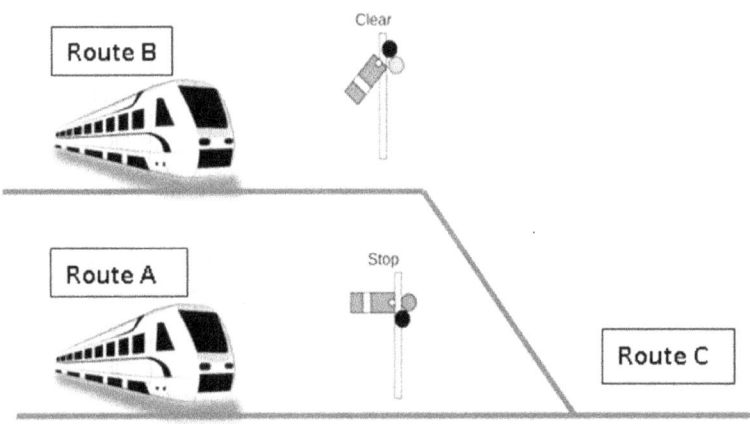

The signalling system will allow a signaller to set a route for the Route B train to proceed onto Route C. The Route A train will be held at the signal until the signalling interlocking system detects that the Route B train has passed the next signal along Route C, and is safely ahead. This simple junction interlocking system is easy to simulate using a computer today. In the 19th century, the same principle of junction control was used, but the interlocking for semaphore type signals was achieved mechanically. Early in the 20th century, electrical relays provided the interlocking logic systems, and in the 1980s, the first solid state interlockings were introduced. Most modern signalling systems in the 21st century utilise computer based interlockings.

If we now extend the basic junction train movement control system model principle to any of the eleven London terminus stations, it will help you to understand why it is crucially important to be able to carry out tests to simulate partial system failure. Being able to manage perturbations in railway timetables, and to recover from partial system failure, is essential to the reputation and credibility of current day train operating companies. Testing recovery strategies, and using the results to instruct and train railway movement controllers and managers, helps to get many thousands of commuters into city centres and back home safely, and on time each day.

Getting Acceptable Results

At the start of this chapter, I extolled the virtues of devising metrics for your system. A basic part of this is to decide what can be measured in your system, what needs to be measured to ensure that the system outputs achieve what the objective or goals are for your system, and what results would be acceptable.

Back in Chapter 1, I noted the use of SMART objectives for your system, where the mnemonic acronym SMART meant:

Specific
An identifiable area of the system which has a specific requirement associated with it

Measurable
A part or area of the system that is able to be measured, or at least provide an indicator of progress towards a

bigger objective

Achievable
Something which can be attained as an identifiable element of the system, or which can be assigned as the responsibility of an individual or team

Realistic
Identify what results can be realistically achieved given the resources available, and is relevant to the system

Time-related
Specify when the required result or output can be achieved by the systemWhere systems are more complex and have subsystems with interfaces inside the overall system, it may be helpful to have a testing strategy that includes unit and subsystem testing, so that the interfaces between units and subsystems can be checked and verified individually. This helps to reduce the time for fault-finding if the overall system output does not reach acceptable levels.

Systems that have elements of tailored or bespoke software in them will benefit particularly from specific testing of identifiable coherent parts; for example, algorithms, processing, and data storage and retrieval elements. Robust configuration control is also important where access to source code is easy, so that installed versions of operating software can be reliably recorded during system testing.

Collecting data will need management attention: with digital technology, it may be straightforward to log substantial quantities of data, but it may be useless if there is so much that spotting faults and trends during testing becomes impossible.

Accelerated whole life testing may be necessary to provide data to support marketing decisions about duration of warranty or guarantees for your system. Getting acceptable test results will depend largely on the type of system and its intended use and output. However, one fact is common across most systems, and that is that time and effort is needed for testing and commissioning if your system is going to be a success.

In the next chapter, we will examine the foundation of success, and that is system assurance.

There is a website for this book:
www.sensationalsystemsbook.com

This provides access to a summary of the tips given in this book for successful system delivery. It will also give access to a workbook to help you develop the techniques on which this book is based.

The author, Geoff Miles, is a Chartered Engineer, and runs a coaching, mentoring, and consultancy business. He is also available for teaching and speaking engagements.

Details can be found at:
www.geoffmilesconsulting.com

Chapter 8

System Assurance

"In our reasonings concerning matters of fact, there are all imaginable degrees of assurance, from the highest certainty to the lowest species of moral evidence. A wise man, therefore, proportions his belief to the evidence."
– David Hume, philosopher, historian, economist, and essayist

This chapter is about the part that system assurance has to play in ensuring that your system is successfully delivered and is sensational at producing the output it was intended and designed to provide.

Creating a V Life-cycle Model for Your System

We need to be clear about what we mean when reviewing System Assurance. As with most things in life, there are fashions and trends that give rise to jargon words and phrases used to set apart the *new* thing from that which has gone before. System Assurance is no exception to this. We used to have inspectors and testers, then we had quality control and quality assurance, and now we have technical assurance and technical *cases*, and so on. For the purposes of this book, I will go back to the basics of what we are trying to achieve with our system. Fundamentally, we want to do two things: firstly, to *verify* that what we set out to deliver as a system is actually delivered, and secondly, to *validate* that what our client wants from the system is actually provided.

To understand this verification and validation process more clearly, I will use the V life-cycle model. Different organisations, industries, and market sectors use the V life-cycle model in different ways and with detailed stages or steps in the process, but generically, the principles are the same. The V-model is a graphical representation of a system development life-cycle, and is used to produce specific system development life-cycle models and project management models.

The V-model summarises the main steps to be taken in conjunction with the corresponding deliverables. It describes the activities to be performed and the results that have to be produced during product development.

The left side of the V represents the development of requirements and creation of system specifications. The right side of the V represents production and integration of parts and their verification and validation. Verification is establishing that what is being built satisfies the requirements, and validation is establishing that what is being built satisfies the objectives (i.e. the user needs). Validation can be expressed as, "Are you building the right thing?" and verification by, "Are you building it right?"

A generic version of the V-model is given below.

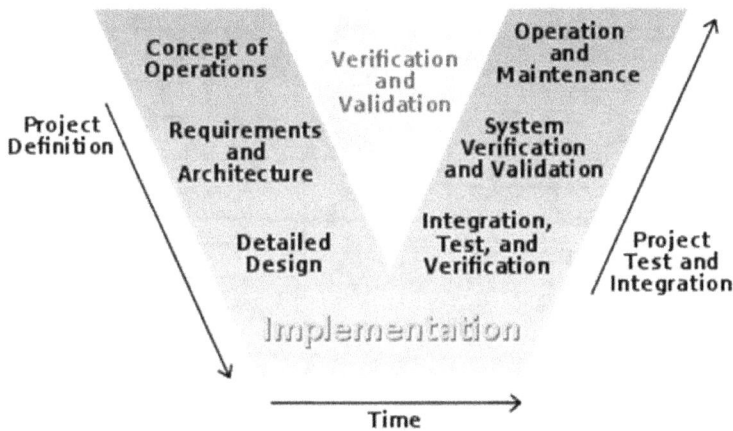

Coming from a railway engineering background, I tend to use the 8-stage V-model because of its widespread adoption across a range of systems.

1. Concept
2. Feasibility
3. Option selection
4. Outline design
5. Detailed design
6. Building and construction
7. Testing and commissioning
8. Acceptance and handover

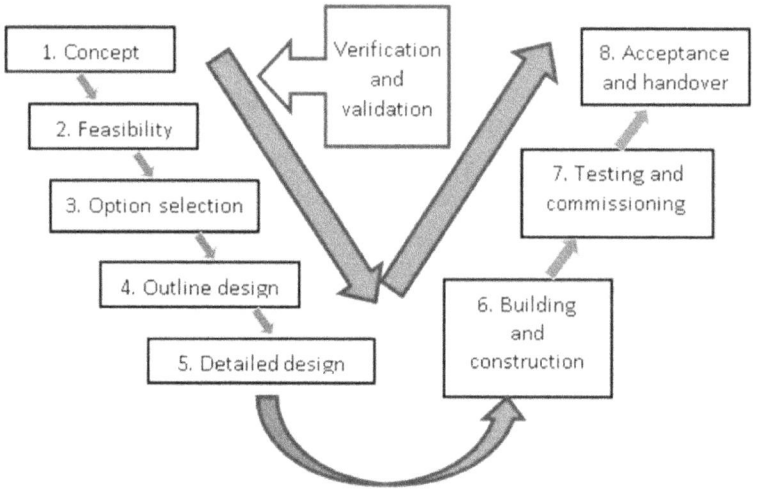

This model can be adapted for use with your system.

The Assurance Plan

An Assurance Plan will be needed for all but the simplest of systems. Assurance is a big word because it has a myriad of meanings, but for the purposes of delivering sensational systems, I suggest that three types of assurances are needed:

- Technical
- Safety
- Quality

Each of these three will have verification and validation techniques associated with them, and working on the principle of *getting it right the first time*, we need to plan our requirements based on the V-model stages. Typically, in complex systems, there will be specific plans produced for each of the types of assurance. This is because there are different disciplines involved with each, with potentially different objectives.

Basic verification techniques are given in numerous codes of practise and industry standards, and include the following:

Inspection—this technique is based on visual or dimensional examination of an element; the verification relies on the human senses or uses simple methods of measurement and handling. Inspection is generally non-destructive, and typically includes the use of sight, hearing, smell, touch, and taste, as well as simple physical manipulation, and mechanical and electrical gauging and measurement. The technique is used to check properties best determined by observation (e.g., paint colour, weight, documentation, and listing of code).

Analysis—this technique is based on analytical evidence obtained without any instrumentation of the system being assured using mathematical or probabilistic calculation, logical reasoning, modelling, or simulation under defined conditions to show theoretical compliance. It is mainly used where testing to realistic conditions cannot be achieved or is not cost-effective.

Demonstration—this technique is used to show correct operation of the subject system, and operational and observable characteristics, without using physical measurements (minimal or no instrumentation or test equipment). It generally uses a set of actions selected to show that the system response to stimuli is suitable, or to show that operators can perform their assigned tasks when using the system. Observations are made, and compared with predetermined or expected responses.

Test—this technique is performed on the system being assured, by which functional, measurable characteristics, operability, supportability, or performance capability is quantitatively verified when subjected to controlled conditions that are real or simulated. Testing often uses special test equipment or instrumentation to obtain accurate quantitative data to be analysed.

Simulation—this technique is performed on models or mock-ups (not on the actual physical system components) for verifying features and performance as designed.

Sampling—this technique is based on verification of characteristics, using samples. The number, tolerance, and other characteristics must be specified and be in agreement with the *glass case* sample (glass case samples are used to establish the benchmark for sampling pass criteria setting).

A typical assurance plan would contain the sections dealing with design, technical, safety, and quality assurance methodologies. Each type of plan would describe what activities would be carried out, who would do it, when they would do it, and how the results would be recorded.

Difficulties sometime occur when the verification criterion used during visual inspection is not easily measurable in a precise way. For example, the contract may say that "the train body sides shall be clean." So, how does the inspector decide whether the train is clean? One person's opinion is likely to be different from the next person.

One way of dealing with subjective assessment for verification purposes is to define specific descriptive words, such as:

Extensive means the area of coverage of the aesthetic defect is greater than 33 per cent but less than 66 per cent of the area being assessed.

Extremely means the area of coverage of the aesthetic defect is equal to or greater than 66 per cent of the area being assessed.

Graffiti means painted, written, sprayed, or scratched graffiti and/or stickers.

Spotlessly Clean means the area being assessed has no evidence of dirt, or streaking caused by poor rinsing, and the finish is not dulled.

Then, using these definitions, a score can be allocated between 0 and 10 to give an assessment metric score.

Sensational Systems

	Condition	Score
A	Spotlessly Clean	10
	Condition falls between A and B	9
B	Very clean appearance, i.e. no dirt noticeable but dull, not shiny	8
	Condition falls between B and C	7
C	Generally clean, i.e. some minor marks, dust, dirt, or streaking	6
	Condition falls between C and D	5
D	Many dirty marks	4
	Condition falls between D and E	3
E	Extensive dirt, including door steps or door controls having inadequate contrast with their surroundings, or visibility of platform signs through windows reduced	2
	Condition falls between E and F	1
F	Extremely dirty, with risk of soiling upon contact, or any stain caused by striking of birds, animals, or humans, or by lubricants, or visibility through windows varies	0

The Assurance Plan for a system is a very important component of the system management armoury because it is able to link the commercial, design, production, operation, and maintenance aspects together into a cohesive group of activities. This book is not specific to any particular system, market area, or industry, and is intended to provide generic information to help in delivering sensational systems. The following text is no exception and uses a typical system assurance plan template to provide an example of what such a document might look like; it will need to be adapted and tailored to suit the particular system that needs to be assured.

This System Assurance (SA) plan template includes four sections:

Introduction—describes the purpose of your SA plan, who was involved in developing it, and how it is and will be used within your organisation.

System Assurance Standards—describes all of your organisation's SA standards across the different content areas (such as design, production, staffing, interventions, client interactions, data, and services).

Dissemination—describes how, when, and to whom (both within and outside your organisation) you will distribute your SA plan and share your findings.

Work plan—a chart to specify SA activities for each of the content areas, and identify the person(s) responsible for each activity, and a timeframe for implementation.

Introduction – Write a brief section that answers the following questions:

- What is the purpose of your SA plan?
- What role will this plan play within your organisation?
- Who was involved in developing the plan?
- What was the process for developing the plan?

System Assurance Standards – This is the heart of the SA plan. This section should be as long as necessary to relate the system requirements to the standards that need to be achieved in system production.

Dissemination – Write a section describing how, when, and to whom you will distribute your SA plan. Consider staff within your organisation, as well as funders, partners, users, and other key stakeholders. Also include how, when, and with whom you will share the findings from your SA plan.

Work plan – The work plan section of your SA plan specifies individual SA activities for verification of the detailed requirements for each of the content areas, and identifies the person(s) responsible for each activity, and a timeframe for conducting the verification activity.

Collecting Evidence

The standard ISO/IEC/IEEE 15288 has two important definitions of the purposes of verification and validation activities:

The purpose of the verification process is to provide objective evidence that a system or system element fulfils its specified requirements and characteristics.

The purpose of the validation process is to provide objective evidence that the system, when in use, fulfils its business or mission objectives and stakeholder requirements, achieving its intended use in its intended operational environment.

To deliver against these purposes, assurance evidence needs to be collected. Depending on the type of verification and validation activity, evidence can be collected as electronic data, photographs, written reports, standard forms, and so on.

We'll deal with progressive assurance in the next section, but for collected evidence, it is important to have a clear way of linking that evidence to the particular system being assured. There are a number of ways of doing this; most are normally based on proprietary software. The software used for information collation will depend on the type and complexity of the system being assured. Historically, evidence was based on the drawing numbers for the components, sub-assemblies, assemblies, and the general arrangement of the system. With the widespread use of Computer Aided Design (CAD) systems, since the 1990s, and their expansion into requirements management systems, bills of materials generation, and full enterprise resource planning, software to integrate system production processes, including assurance, evolved.

A step along the digital road was the introduction of Building Information Modelling (BIM). The Handbook of BIM (Eastman, Teicholz, Sacks & Liston 2011) defines BIM as:

With BIM technology, one or more accurate virtual models of a building are constructed digitally. They support design through its phases, allowing better analysis and control than manual processes. When completed, these computer-generated models contain precise geometry and data needed to support the construction, fabrication, and procurement activities through which the building is realised.

It wasn't long ago that Enterprise Resource Planning (ERP) envisioned a single system to manage and unify all of a company's business needs. It was the promise of a unified system of integrated applications to manage the business and automate the back-office functions relating to technology, services, and human resources that has driven the development. Even this is being superseded by Enterprise Resource Platform. This will serve as an organisational platform, delivering cross-functional capabilities of flexibility, openness, mobility, collaboration tools, user experience, artificial intelligence, and use of the digital cloud.

Verification and validation require assurance checks to be carried out against defined criteria, and procedures need to be in place to deal with failure, to achieve the level set for pass success. In larger companies, Non-Conformance Report (NCR) procedures are an established part of the quality management system. Copies of NCR documents are placed in the assurance files, along with all the other assurance documentation. Close-out of defects is recorded in additional parts of the NCR document so that evidence can be collected for the final acceptance of the system, showing its full compliance with requirements.

It can be seen that whatever size of system you need to assure, you will need to establish how assurance evidence, which you will collect, can be related to the version of the drawing that you are verifying, and the configuration of any related software components held within the system at the time of the evidence gathering.

Progressive Assurance

It used to be that quality assurance was a process that took place at the end of the development cycle, as if quality was a tangible feature that you could add to the product before shipping. When defects were discovered, it took a major effort to identify the root causes and fix the problems. In today's world of increasingly complex software-driven smart products, quality must become an integral part of the systems development process for there to be any hope of delivering products that are not only defect-free but demonstrate a fitness for purpose. In this section, we take a look at how systems assurance helps identify any issues, through validation and verification early in the development cycle, leading to improved chances for system success.

Earlier in this chapter, we looked at the eight stages of a typical system V life-cycle:

1. Concept
2. Feasibility
3. Option selection
4. Outline design
5. Detailed design
6. Building and construction
7. Testing and commissioning
8. Acceptance and handover

If all assurance activity was left to the last stage, it is likely to introduce unquantifiable additional system cost and delay. The way to avoid this is to carry out progressive assurance (i.e. verify and validate your system throughout its development life-cycle, which should be maintained during every stage of system development).

Alternatively, you could select the critical stages of the life-cycle. These will generally be at the starting point at which the development starts to incur significant expenditure. Obviously, depending on the type of system, significant expenditure will commence after stage 2 (establishing the feasibility of your system and starting to review designs for options for the way forward, typically 10% of total cost); after stage 4 (completion of outline design and committing to detailed design and production, typically 80% of total cost), and of course, at stage 8, acceptance and handover (at the end of which, the client should pay you for completion of the system). Carrying out Stage Gate Reviews is discussed in the next section.

Your System Assurance Plan should include a matrix that shows the eight life-cycle stages across the top X axis, and the required assurance documentation on the vertical Y axis. The matrix is then populated to show which documents are required at which stage during the development life-cycle.

System Assurance

VERIFICATION AND VALIDATION MATRIX EXAMPLE	1. Concept	2. Feasibility	3. Option selection	4. Outline design	5. Detailed design	6. Building & construction	7. Test & commissioning	8. Acceptance & handover
Drawing approval		✓	✓	✓	✓			✓
Risk assessment	✓	✓	✓				✓	✓
Testing procedure				✓		✓	✓	
Conformance report				✓			✓	✓
Certification							✓	✓
Acceptance procedure				✓			✓	✓

This is a very simple example; the number of different documents to be produced as part of your system progressive assurance plan will be tailored to meet your own system needs.

Stage Sign-off

As noted in the previous section, the V life-cycle for your system should have a number of distinct stages. It makes good management sense to acknowledge the completion of one stage before proceeding onto the next one. One way of formalising the stage sign-off is to conduct a Stage Gate Review.

A Stage Gate Review (SGR) is a progress meeting between the sponsor (or client) and the System Project Manager, whereupon the health of a project is checked before progressing on to the next stage within the life-cycle, or requesting additional investment authorisation. All SGRs should be planned at the system development outset, and included in the development plan and the System Definition Document within the System Management Plan.

Documentation for the entire system development project should be identified at the start of the system development, in a matrix similar to that developed for verification and validation documentation, to form the Stage Gate Checklist.

This will be particularly important for complex systems (i.e. those that contain systems within systems). It may be that one system will need to be at stage 8 (acceptance) before it can be built in to the larger system, which may only be at stage 6 (building and construction). As an example, Network Rail (the UK's infrastructure owning, maintaining, and renewing organisation) developed a process containing over 150 different documents, which needed to be in place to fulfil all 8 of their stage gate checklists. SGRs are undertaken once a Stage Gate Checklist has been completed, and documentation prepared, for review at the stage gate meeting.

It is the completion and authentication of the documents detailed on the particular stage gate checklist that is the secret of success for system development. Diligent validation that the work on system development has been carried out in accordance with sponsor requirements, standards, and other compliance requirements by the system project manager, prior to the presentation of the information and documentation to the relevant stakeholders, will ensure that you *get it right first time*.

In order for the system financiers to have confidence in giving approval for further expenditure, the completed Stage Gate Checklist should accompany any investment authorisation documents. In that way, the necessary evidence is available in support of the current system status when applying for investment authority.

The actual SGR meeting should be very much a formality when the attendees (including relevant stakeholders and the business directors) have trust and confidence in the system project manager. If they have procedures in place that utilise the principles of system development techniques using the V life-cycle model, then sensational system assurance can be guaranteed.

In summary of this section, verification is intended to check that a system (or a part of it) meets a set of design requirements specifications. In the development phase, verification procedures involve carrying out special tests to model or simulate a part, or the entirety, of a system, then performing a review or analysis of the modelling results. In the post-development phase, verification procedures involve regularly repeating tests devised specifically to ensure that the system continues to meet the initial design requirements, specifications, and regulations, as time progresses. It is a process that is used to evaluate whether a system complies with regulations, specifications, or conditions imposed at the start of a development phase. Verification can be carried out in development or production stages, and is normally an internal process.

Validation is intended to ensure that a system (or a part of it) meets the operational needs of the user. For a new system development, validation procedures may involve modelling and using simulations to predict faults or gaps that might lead to invalid or incomplete verification or development of a system. A set of validation requirements (as defined by the client, sponsor, or user), specifications, and regulations may then be used as a basis for qualifying a development or verification for a system. Validation procedures also include those that are designed specifically to ensure that modifications made to an existing qualified system, or a new or replacement system, meets the initial design requirements, specifications, and regulations.

It is the process of establishing evidence that provides a high degree of assurance that a system accomplishes its intended requirements. This involves acceptance of fitness for purpose with end users and other system stakeholders. This is often an external process.

It is sometimes said that validation can be expressed by the query, "Are you building the right thing?" And verification by, "Are you building it right?" *Building the right thing* refers back to the user's needs, while *building it right* checks that the specifications are correctly implemented by the system. In some contexts, it is required to have written requirements for both, as well as formal procedures or protocols for determining compliance.

It is entirely possible that a system passes when verified but fails when validated. As an example, this can happen when a system is built as per the specifications, but the specifications themselves fail to address the user's needs. After all the hard work of getting the system up and running, the next chapter gives some respite by looking at operation and maintenance.

**There is a website for this book:
www.sensationalsystemsbook.com**

This provides access to a summary of the tips given in this book for successful system delivery. It will also give access to a workbook to help you develop the techniques on which this book is based.

The author, Geoff Miles, is a Chartered Engineer, and runs a coaching, mentoring, and consultancy business. He is also available for teaching and speaking engagements.

**Details can be found at:
www.geoffmilesconsulting.com**

Chapter 9

Operation and Maintenance

"Look at a day when you are supremely satisfied at the end. It's not a day when you lounge around doing nothing; it's when you've had everything to do, and you've done it."
– Margaret Thatcher, British Prime Minister, 1979 to 1990

This chapter is about what is needed to get your system into operation, and once there, what is needed to maintain it to keep it in successful operation.

The Training Plan

Back in Chapter 3, we considered who the stakeholders for our system might be, and we looked at the different involvement of owners, users, operators, and maintainers, once the system is brought in to use. Generally, for any system, some form of training will be needed, if only to ensure safety of the people who will use it.

Our Boiled Egg for Breakfast (BEFB) system has inherent safety risks for the user because it involves the use of boiling water and high temperatures to heat the water. If we were introducing this system to one of our children so that they could cook their own breakfast, as a good parent, we would inherently warn them of the dangers and the risks of being scalded or burnt. We would ask that they take precautions, such as using gloves, ensuring that saucepans do not sit on the stove with handles sticking outwards, and so on. We could write this out as part of a structured training plan for the BEFB system. The generic content of a training plan could be as follows:

- System name
- Who is to be trained
- Training objectives, safety, and environment considerations
- Risks and hazards
- Training structure and session plans
- Facilities and materials required for training
- Activities and learning strategies
- Training schedule and activity durations

For more complex, large scale systems, such as a railway electrification system, training plans are likely to be templated by the system sponsor or client, and may include requirements driven by legislation. Publicly funded systems will have contractual conditions associated with training and employment. You may need to consider such factors as:

- Creating apprenticeships
- Employing a percentage of graduates
- Employing people from local areas
- Ethnicity of staff

- Clear evidence that you do not discriminate against age, sex, or colour

All of these will need to be reflected in your training plans by creating training modules for various groups and academic capabilities. Thought must be given to who will do the training: will it be out-sourced to specialist companies, or do you need to develop in-house training capabilities?

Where your systems will be used by clients, with many staff to be trained, system providers may use *train-the-trainers* plans. This is where the supplier trains the trainers from the client company, who then develop their own bespoke training plans for their employees.

Establishing and checking system user competence may be required where systems are installed in safety-critical applications. Aircraft are an obvious example of this. Airline pilots need to undergo extensive training prior to initial certification, and then, subsequently, have their ability to deal with various fault scenarios checked on a regular basis. Individual training plans, coupled with log-books, are essential components of the pilot's license, to fly both private and commercial aircraft.

The use of simulators in training plans is widespread. Train drivers need route knowledge education and competency assessment, and any one of the 26 franchise service train operators in the UK will have upwards of 400 drivers on their staff roster lists. Driving simulators form a critical part of economically providing such facilities as route learning and how to deal with train faults and obstacles on the track. Car, truck, and bus driving training and assessment follow similar principles.

Where systems (or parts of them) are designed and manufactured outside of the developer's country of origin, there may be trade agreements between the countries involved that requires *local content*. For example, resource-rich countries in the Middle East and North Africa (MENA) region, especially Gulf countries—Kuwait, Iran, Iraq, Bahrain, Oman, Qatar, Saudi Arabia, and the United Arab Emirates (UAE)—are increasingly inserting local content requirements (LCRs) into their legal framework, through legislation, regulations, guidelines, industry contracts, and bidding practices. Despite the clear and uniform overall policy driving the use of LCRs and product mandating requirements in the MENA region, approaches taken to enforce LCRs vary, and must be carefully understood and clarified to avoid misalignments and contractual mismatch between governments and international companies.

System Operating Regime

At the concept stage of system development, thought must be given to the operating regime for the proposed system. In simple terms, this can be concerned with who will use the system, and where and when will it be used; and the "why" question should be answered by the output that the system will provide.

Back in Chapter 1, we looked at creating a picture of your system. The picture will help to set the basic requirements for the operating regime; for instance, in terms of physical size, whether it is portable or static, located indoors or outside, required physical robustness, domestic or industrial use, permitted noise levels, and so on. It is important to know who will operate the system, mainly because of the ergonomics that will need to be designed into it, and the amount of automation to be provided in the operating systems. Railway signalling provides a good example of how system design has changed with the progression of the operating regimes of modern rail networks. Moving from one track to another is carried out by moving a track switch to the left or the right. In the 19th century, this was done by pulling a lever on the side of the track, beside the switch rails (interestingly, this is still done on lightly used tracks, such as sidings and shunting yards, today). As the railways grew, and track switches in a particular location, such as stations with multiple platforms, increased in quantity, signal boxes were introduced, where all of the levers for a junction were accumulated in one place, with cables going out from the signal box to the various switches and their associated semaphore signals in the vicinity (still in use today at the busy junction in Shrewsbury, Shropshire, UK).

The system used to ensure that train routes that conflicted with each other could not be inadvertently set by the signaller, was in the form of mechanical interlocking between the switch levers, to physically prevent opposing routes to be set, which might allow trains to approach each other on the same tracks. The system operating regime was one where there was a competent operator (the trained signaller), but that person was doing regular monotonous work, and potentially, mistakes could easily be made, with catastrophic consequences.

During the 20th century, with rapid electrical technological progress, semaphore signalling with mechanical interlocking was phased out and slowly replaced with electrical route relay interlocking (RRI) technology. This, coupled with electrical position detection of track switches and electro-pneumatic and electro-hydraulic switch motor actuation, gradually saw the phasing out of the ubiquitous signal boxes (where the control levers were located in a room above the big mechanical lever frame counter-balance weights, pulleys, and mechanical interlockings). These were replaced with local area control rooms.

In the 1980s, at the start of industrial application development of computers, solid state interlockings (SSI) technology started to replace the RRI systems. Although more trains were being run on the UK rail network, and track junctions became more complex, the SSI technology was able to reduce the number of local control rooms, and incorporate them into large area control centres, known as Integrated Electronic Control Centres (IECCs).

On into the 21st century, and the start of the digital age, SSIs are being phased out, and these are being replaced with computer based interlockings (CBI) technology. The 14,000 route miles of the UK national rail network, with some 1400 local area interlockings, have the main intercity routes all controlled from just twelve Route Operating Centres (ROCs). This is leading to the elimination of over 800 signal boxes, and the removal of around 200 local

Operation and Maintenance

control signalling panels and IECCs.

The railway signalling example serves to show how the basic system operating regime is largely unchanged, as people are still using a means of interlocking the route-setting track switches to prevent collisions between trains. However, the method of delivering the output has changed, with the change in technologies capable of providing the system with high levels of reliability and safety.

Having computers and associated system software carry out some of the tasks previously done by humans has had a major impact on operating regimes. A major benefit has been reduction in costs, and improvement in the consistency and associated quality of products. From the point of view of the human factor, people have been replaced by machines for monotonous and dangerous work; however, this has the significant disadvantage of reducing the number of jobs for manual and semi-skilled people. Systems are the winners in this operating regime scenario.

Apart from systems that have a series of physical component parts designed to achieve a defined output, another large group of systems, where the operating regime is important, are those concerned with a specific process. The component parts of those systems may not all have a physical form but are to do with a change of state or manipulation of material or data.

The development of mathematical models is a fairly recent feature in the application of advanced model-based techniques for control optimisation, scheduling automatic fault detection, and diagnosis in the process systems industries. Hence, there is a potential for improved system product quality, as well as improved production system flexibility and safety. The technique includes non-linear modelling and identification using a combination of empirical system process data and prior knowledge.

The mathematical modelling techniques employed are beyond the scope of this book, but the principles include establishing a modelling framework based on operating regime decomposition within the system's operating range. Within each operating regime, the system is modelled with a simple local model. These local models are then patched together using a smooth interpolation technique. This framework supports the development of complex system models and is flexible with respect to the amount of prior knowledge and empirical data required.

System Maintenance Regime

Unless they are *single-use* systems, the majority of systems will require some degree of maintenance. Even the BEFB system will need the pan, crockery, and utensils cleaned occasionally. As part of the development of a system, consideration should be given to the frequency, extent, and complexity of the maintenance tasks. Furthermore, the design must allow for the maintenance activities to be carried out in a cost-effective way. For example, if the oil level in the gearbox needs to be checked weekly, then a sight glass or dipstick should be provided in an easily accessible location.

Operation and Maintenance

It is interesting to think of the changes that have happened to car engines over the last fifty years. As previously stated, some maintenance manuals from cars built in the 1960s were telling us to check the engine oil level weekly, and change it every 5000 miles. With improvements to engine technology minimising the amount of oil burnt by the engine, and improvements to the lubrication capabilities of synthetic oil products, modern engines are capable of running for 12,000 miles or more without loss of oil, or the need for it to be changed. A glance at the engine compartment of a well-designed modern car will quickly show the locations for fluid checks and top-ups, by colour coding and large caps capable of being removed without any tools other than a human hand. It is helpful to think of maintainability when designing the maintenance regime for your system.

Many systems contain software, which may, itself, need to be maintained to incorporate new and upgraded features, as well as to eliminate defects and bugs in the software coding. Display screens may be built into the system, either as part of the system controls or as a stand-alone diagnostic feature. Smart phones are an obvious example of how software can have built-in maintenance regime features; the use of Apps takes this to a whole new level. However, the suppliers of software do not always get it right; Microsoft is a prime example where software upgrades can take you by surprise, by taking over the whole computer operating system quite unexpectedly, and in a very user un-friendly manner. A further problem comes with upwards compatibility, where add-ons, like printer drivers, do not work with version upgrades of operating software. Think about the user for your system software when carrying out maintenance.

For complex systems, or systems within systems, there needs to be a *controlling mind* for the whole set-up, so far as maintenance is concerned. Just think about road surface maintenance; how many times do we see a replacement road surface laid for the first time in ten years, after having suffered pot-holes for the last five years, only to see the water company come along a month later and dig a hole in the road to fiddle with something underground. Worse still, the gas company comes along a couple of months later and digs another hole not far from the water company's hole. The result is another ten years of rough road, and further potholes. If the local highway engineer acted as the controlling mind for the road systems, inconvenience for the user could be avoided.

Two further horror stories to end this section, from the author's experience, concerns building services systems. The first concerns a tap installed on an internal wall in a workshop, where water is needed for filling car screen wash fluid and for small parts cleaning. The drains in the workshop were installed by a civil engineering contractor, and the taps were installed by a building services contractor. The obvious result was that the tap was not sited over the drain, so that when the hose was disconnected from the tap, water was spilled when a bucket was being filled, the water puddled on the workshop floor, and caused a slip hazard. Both contractors denied responsibility for the error, and extra money had to be paid (to move the tap) to correct the problem. One of the lessons learnt was to ensure that all interfaces between systems are clearly defined, and responsibility for managing them unequivocally allocated.

The second example (in the same building but with different contractors) concerned the building climate control and fire suppression systems. A large air handling unit, with fans and filters, delivered fresh air into the building from outside, into the internal air conditioning system. The unit was located in its own plant room. The fire safety engineers designed a fire detection and suppression system for the plant room, and installed the sprinkler pipework around the air handling unit. One of the water supply pipes was routed along the side of the ducting, which contained the air filters. The filters needed to be changed every few months, so it wasn't until the first scheduled filter maintenance session that the engineer found that the filter could not be withdrawn from the ducting because it fouled on the fire system pipework. A costly re-routing of the fire pipework had to done and then recertified by the fire authorities to correct this.

It is important to pay diligent attention to system maintenance regimes if you want to deliver a sensational system.

Operation and Maintenance Documentation

Simple systems, like domestic appliances, require operation and maintenance documentation because of legislation that is concerned with user's rights and health and safety. Even if legislation did not apply, it is good practise for system developers to provide information about how best to use the system, how to maintain it, how to identify problems, and what to do if problems occur. Inevitably then, operation and maintenance (O&M) documentation will be needed for your system.

The word, *documentation*, used to mean hard copy paper, leaflets, books, lever arch files, and so on, but with the wide-spread use of the internet, particularly in the developed world, O&M documentation is now more likely to be something that you look up on a website. You only download the part that is relevant to you and your use of a system. Smart phones are a prime example of how paper documentation has all but disappeared: you get a note to tell you how to turn it on, and a few basic uses; then the rest is up to the user. How many people, who have paid a premium price for it, know the true extent of the awesome power built in to Apple's iPhone X?

For more conventional bespoke systems that have been designed and built to deliver a specific outcome for a paying client, O&M documentation will be required, regardless of the form it will take. Therefore, time and cost need to be allowed for the effort needed to produce what is required as part of the resource planning and scheduling for the system production.

Maintenance requirements can be established during the design stage of a system. Simple analysis is done, using questions such as:

Operation and Maintenance

- What can wear out?
- What is its estimated life?
- How often does it need to be replaced?
- How will replacement be carried out?
- Where will replacements come from?
- Which items might suffer from accidental damage?
- What test equipment might be needed?
- Are any special tools required?
- Is training needed for maintainers?

These questions, and those like them, should provide the starting point for O&M documentation. Where the system is produced using engineering drawings, a set of approved as-built drawings can be used to provide the basis for exploded views of assemblies, to show how the component parts are put together. The Computer Aided Design (CAD) produced as-built drawings can be used to compile detailed parts lists, with unique part numbers and quantities required of each. Larger companies may use technical writers to support the drawings, with descriptive text for carrying out maintenance. Specialist companies are able to do the whole job of O&M documentation production under contract.

Presentation of O&M material may be important when used to promote the business that originated the system, and it should be ensured that branding is incorporated throughout. It can be web-based or distributed on DVDs or flash drives. Where intellectual property protection needs to be considered, use of patents and other copyright legal mechanisms may be needed. Typically, protecting bespoke software used in a system gives rise to a minefield of legal difficulties. The client who has commissioned the development of a system using bespoke software will want to protect their investment by receiving a copy of the source code for the software. This will be in case the software supplier goes out of business, and the system software needs to be changed or modified. One way of dealing with this is for a copy of the as-delivered software source code to be held in escrow by an independent legal third party, who will be given instructions by both parties about the terms for releasing the data.

At the time of delivery and handover of a bespoke system containing software, the configuration of the system, and the version of any software installed, will be agreed. Any as-built drawings will be listed together with the version number of each drawing. Commencement of any period of warranty, defect correction, or free servicing is a key point in a system's life-cycle, so comprehensive and accurate documentation at that time is essential.

Where systems are sold outside the national boundaries of the country in which they are manufactured, attention has to be given to the language used in documentation. I'm sure we all know of systems that we have bought, where the O&M documentation comes with 100 pages, but because the system is sold in ten different countries, only 10 pages are relevant to us.

Asset Management

All systems will contain various components, elements, items, sub-assemblies, parts, etc., collectively called assets.

Operation and Maintenance

Assets, and the value realised from them, are the basis for any organisation delivering what it aims to do. Whether public or private sector, or whether the assets are physical, financial, human, or *intangible*, it is good asset management that maximises value-for-money and satisfaction of stakeholders' expectations. It involves the coordinated and optimised planning, selection, acquisition, development, utilisation, care, maintenance, and ultimate disposal or renewal of the appropriate assets and asset systems.

Insights into the integration and optimisation of asset management have developed since the 1990s (from the North Sea oil and gas industry, and the Australian public sector), to identify a range of essential business processes, alignment activities, and system integration features that yield very significant performance benefits.

The first task in asset management is to develop an asset management plan (AMP). The AMP will start from a register of assets that go to make up your system. The asset register will derive from the lists of materials identified from the system design. Each item in the asset register will need a unique identification number. There are numerous books that deal with the complexities of part numbering systems alone, but for our purposes, you may find it helpful to have some intelligence incorporated in the number. For example, in the system breakdown structure, assemblies at level 1 start with the number 1; those at level 2 start with 2, and so on. The number needs to be long enough to include a version number or letter. Using an alpha-numeric numbering system may give more scope for intelligence.

A problem that some larger systems encounter is that suppliers of sub-systems insist on using their own part numbers in all documentation. To avoid having to edit their component documentation to make it fit with your requirements, it is possible to allocate meta-data (metadata is data [information] that provides information about other data; it is used to summarise basic information about data, which can make tracking and working with specific data easier), and use automated computer-based look-up tables to link their numbers to your methodology. Having compiled a comprehensive asset register for the components of your system, it is then possible to create a matrix of such associated data as:

- description and specification
- drawing reference number
- level of maintenance required (if any) and how often
- whether the component must be replaced or is repairable
- time required to replace or repair
- level and type of skill required for maintenance

Although many systems will be unique or substantially different from each other, there are a number of generic asset management software systems on the market that are designed to be easily tailored, to be able to carry out specific systems asset management. Also, they are increasingly being made a part of a total enterprise management suite of software so that they are able to take actions such as absorb data from remote condition monitoring (RCM) systems, determine when maintenance needs to be done, issue works orders for the work, raise purchase orders to buy spare parts, and to update system maintenance records automatically.

Asset management is rapidly becoming a science in itself, and standards have been produced and established for companies to demonstrate their competence in the discipline by certification. This is done by the International Organisation for Standardisation (ISO). The ISO 55000 series provides terminology, requirements, and guidance for implementing, maintaining, and improving an effective asset management system.

The three international standards (ISO 55000, 55001, and 55002) are important because they represent a global consensus on asset management and what it can do to increase value generated by all organisations.

It is likely that asset management will increasingly become the best means of harnessing and integrating all the activities within an organisation, to achieve its purpose most effectively and economically.

ISO 55000 consists of three standards:

ISO 55000 Asset Management – Overview, principles, and terminology

Introduces the critical concepts and terminology needed to develop a long-term plan that incorporates an organisation's mission, values, objectives, business policies, and stakeholder requirements.

ISO 55001 Asset Management – Requirements
Specifies the requirements for the establishment, implementation, maintenance, and improvement of an asset management system.

ISO 55002 Asset Management – Guidelines on the application of ISO 55001
Provides guidance for the application of an asset management system, in accordance with the requirements of ISO 55001.

These standards can help organisations of all sizes and sectors to:

- Establish an asset management system to optimally manage assets
- Implement, maintain, and improve an asset management system
- Comply with asset management policy and strategy
- Demonstrate that they are applying best practice
- Seek external certification of their asset management system, or make a self-declaration of compliance

**There is a website for this book:
www.sensationalsystemsbook.com**

This provides access to a summary of the tips given in this book for successful system delivery. It will also give access to a workbook to help you develop the techniques on which this book is based.

The author, Geoff Miles, is a Chartered Engineer, and runs a coaching, mentoring, and consultancy business. He is also available for teaching and speaking engagements.

**Details can be found at:
www.geoffmilesconsulting.com**

Chapter 10

Using Your System

"Only those who will risk going too far can possibly find out how far one can go."
— **T.S. Eliot,** poet, essayist, publisher, playwright, literary and social critic

This chapter gives advice on the things to think about when your system is built and is in operational use.

Performance

Performance of your system should have been checked against specified requirements in the system development and production stages. This applies to both single bespoke systems and to volume-produced systems. The system performance can affect not only repeat sales to existing clients but also the number and cost of any warranty claims. Most importantly, it can also affect your company reputation and the strength of any *brand* that you may have.

Branding is a huge subject, and there is plenty of further reading available on the topic. Two points are worth noting:

firstly, your *brand* is worth something, and secondly, if selling sensational systems is the way you grow your business, then you should have a Brand Plan.

Brand equity and brand value are measures that estimate how much a brand is worth. The difference between the two is that brand value refers to the financial asset that the company records on its balance sheet, while brand equity refers to the importance of the brand to a customer of the company. The Brand Plan could be your way of growing the worth of both these measures. The five basic questions to ask are:

1. Where are we? Carry out a situation analysis for your business.
2. Why are we here? Identify the key issues that have led to the business being where it is now.
3. Where could we be? Prepare your vision for the company, write a simple mission statement, and identify your goals for 1, 3, and 5 years, looking forward.
4. How can we get there? Determine your strategy and tactics that you can use to achieve your vision, mission, and goals.
5. What do we need to do to get ready? Establish how you will execute your strategies and tactics, and select how you will measure your progress.

You can see that your system performance is a key element for a successful business. Checking how the system is performing against the specified requirements can be carried out using some of the assurance techniques identified in Chapter 8. These include measurement, observation, and feedback from users. For example, are the inputs simple to provide, or are they costing too much? Are the design outputs being achieved, or are they failing to deliver expectations? Is maintenance simple to carry out, or is it too frequent or costing too much in replacement parts? Are the users and operators comfortable with the system processes, or have they suffered accidents or produced system faults and errors?

Gathering and analysing the evidence on system performance will assist in development of new models of the system, adding value with design improvements for new models, and upgrades to existing models.

Upgrades and Obsolescence

How many of us have been frustrated by our computers, tablets, and smart phones when they are occupied by performing software updates, which takes away their ability to do the things that we want them to do?? It appears to be accepted practise for the owners of the software systems that dictate modern life, to carry out these upgrades with monotonous regularity. Why do they do it? Whilst the excuses vary, the main ones are bug fixes and incorporation of *new and improved* features.

The history of Microsoft Windows® shows how obsolescence can follow a similar pattern as software upgrades. There is some upward compatibility but certainly not all. The significant change up to Windows 10 (following as it did from Windows 7, Windows 8, [and what happened to 9], Vista, XP and the original Windows) caused many problems with associated software, such as peripheral device drivers that just could not be run in the new Windows versions, causing much angst among users. Whilst your business may not have the massive influence of the Microsoft empire, you can learn from their treatment of users should you need to upgrade or replace obsolescence in your sensational system.

Where systems are complex bespoke designs, particularly where software is used, upgrades are virtually essential. This is not always the fault of the basic software code but due to the data that the code relies on as input to generate the required output. A good example of this is the use of traction power software to control train performance on a railway. Modern traction power systems are designed to allow trains to be driven automatically. This depends on a number of factors, including the alignment profile of the tracks that the train runs on (i.e. where the gradients are, and the curves, which may need reduced speed), where the stations are, and where trains need to stop. All of this is input data that will allow trains to run on a route at optimum speed safely. For a recent UK Southern Region new train fleet, the traction power software was at version 12 before it was authorised to carry passengers. With electromagnetic compatibility (EMC) issues (where at certain locations the power thyristors on the train could disrupt the signalling power system, and potentially cause signalling safety systems to fail) discovered during early service operation, the software had to be upgraded to incorporate frequency band filters. Final sign-off of the traction power system by the client came at version 16.

Using Your System

This led to huge cost and programme overruns for the train systems supplier.

System obsolescence can be due to maintenance spares not being made available by the original equipment manufacturer when a sub-system requires a partial redesign, or it could be that technology has moved on so that the system is wholly or partially inefficient, not being bought by consumers, or just worn out and needs to be replaced. Cars are an obvious example of that. Impressive improvements to internal combustion engine fuel efficiency over the last decade, and recent developments in energy storage, allowing the use of hybrid (internal combustion and electric motor power sources), have led to customer-driven purchases of these new systems, which render the older vehicles largely obsolete.

Stakeholder Satisfaction

The phrase, *the customer is always right*, was originally coined in 1909, by Harry Gordon Selfridge, the founder of Selfridge's department store in London, and is typically used by businesses to convince customers that they will get good service at this company, and convince employees to give customers good service. Providers of sensational systems need to give stakeholders satisfaction, but the adage from 1909 really does not apply in modern society. Alexander Kjerulf produced a Blog for the *Huffington Post*, in 2014, which cited the following story:

One woman who frequently flew on Southwest Airlines was constantly disappointed with every aspect of the company's operation. In fact, she became known as the *Pen Pal*, because after every flight, she wrote in with a complaint. She didn't like the fact that the company didn't assign seats, the absence of a first-class section, not having a meal in flight, Southwest's boarding procedure, the flight attendants' sporty uniforms, or the casual atmosphere.

Her last letter, reciting a litany of complaints, momentarily stumped Southwest's customer relations department. They sent it up to Herb Kelleher's desk (he was CEO of Southwest at the time) with a note that simply said: *This one's yours*.

Kelleher considered this for about sixty seconds, then wrote back: *Dear Mrs. Crabapple, we will miss you. Love, Herb*. Mr. Kjerulf thinks that businesses should abandon the phrase, *the customer is always right*, once and for all, because, ironically, it leads to worse customer service.

His top five reasons are:

1. It makes employees unhappy.
2. It gives abrasive customers an unfair advantage.
3. Some customers are bad for business.
4. It results in worse customer service.
5. Some customers are just plain wrong.

I suppose that whether or not you agree with the old adage depends on what area of the systems market you operate in. If a client has come to you with a requirement for a specific output from the system to be delivered from a particular set of inputs, and you agree a contract to

Using Your System

supply on that basis, then that person or organisation is absolutely right to expect what he is paying for. Where systems are supplied to the domestic consumer market, the consumer has rights enshrined in legislation, designed to protect against bogus suppliers with, for example, unsafe products.

The reputation of the supplier has a value, and it may be in the supplier's interest to ensure that the stakeholder is satisfied with his system. The Rolls Royce car maker is able to charge a premium for the product on the strength of reputation built up over nearly a century of production. Tesla, a relatively new entrant in the North American car market, with a sensational system product incorporating a hybrid power system, ranked only 21 out of 27 car brands, in 2017, in an independent review of owner satisfaction.

Where the system is operating away from a bespoke or mass market, it is possible to use bench-marking as a method of quantifying stakeholder satisfaction and how well your system compares with other similar systems. Bench-marking can be helpful in other ways, especially where system funding may be an issue. Looking at how others achieve a similar objective to your requirements enables costs to be compared, and expenditure justified, on the basis of anticipated benefits.

Integration of developed world city transportation systems is an example where urban conurbation leaders use comparison with other similar sized city populations, to assess how well they are doing with such systems as modal interchanges and through-ticketing. Such comparison findings can be used to support funding grants for improvements.

Perhaps though, the most reliable way of assessing stakeholder satisfaction with your system is to go and ask them.

Assessing Whole Life Benefits

The financial aspects of systems are not covered in any depth by this book, but it is important to understand some of the financial basics. Further information on this vital area can be found at www.sensationalsystemsbook.com. One of the main points to consider is that there will be capital costs and operating costs. The difference between these two is that a system capital cost will be the cost of getting the system through the V life-cycle stages, to the point where it moves in to use. The system operating costs will be those associated with such items as:

- Operators wages
- Consumable costs (fuel, paper, ink cartridges, etc.)
- Utilities (electricity, gas, water, etc.) supply costs
- Maintenance charges (labour and replacement parts)

If we add the initial capital cost to the operating cost for the life of the system, we will get the whole-life cost. Most systems will generate some output, and the output will have a value (revenue), which we can then factor up into the system whole-life value. If we take the cost away from the value generated by the system, we can get a figure for the system whole-life benefit in financial terms.

Using Your System

It may be helpful to aid understanding of these fundamental principles by using the example of the BEFB (Boiled Egg for Breakfast) system.

	Element	Cost (£)
A	Design (1 hour)	90
B	Materials	
	• Saucepan	7
	• Plate	3
	• Egg cup	4
	• Spoon	1
C	Testing and commission	45
D	Acceptance and handover	50
	Capital cost	200

Operating costs will be:

	Element	Cost (£)
A	Operators time (10 minutes at £12 per hour)	2.00
B	Gas	0.05
C	Water	0.05
D	Egg	0.20
	Operating cost per breakfast	2.30

If we assume that the local café charges £3.30 for a boiled egg breakfast, the daily saving by using your own system will be £1 per day. Then, if we assume we have 200 boiled egg breakfasts per year, we can save the capital cost by the end of the first year, and each year, thereafter, we will gain a benefit of £200 from our BEFB system.

This is a small example, but more complex and larger systems will follow the same basic principles.

Learning From Experience

Your system is up and running, and you have discussed its performance and delivery of the required output with the user, operator, and direct stakeholders. They are satisfied with much of it, but there are a few niggles, lingering

Using Your System

doubts, or perhaps suggestions for improvements. These need to be captured into a list of issues, and an action plan developed for dealing with them.

Additional groups of issues will come from two other sources. Firstly, there is feedback from the people responsible for maintaining your system. They might report that a component is difficult to get at, an assembly is failing too often due to vibration fractures of support structures, the belt tension needs frequent adjustment, or the software crashes unexpectedly so that the control system needs rebooting, and so on. All of these need to be added to the issues log. The second group of issues may come from the risk register for the system. Ideally, you will have started the risk and hazard log right back at the concept stage, and continued with it throughout the V life-cycle, up to the point of acceptance and handover to the client.

At that point, you should have closed out all of the risks previously identified in the design, construction, rework, modification, testing, training stages, operating instructions, and so on. Inevitably, there will be residual risks, all of which will be classed as tolerable. Note that if any risks fall into the intolerable category, they must be closed out before final acceptance of the system. You should review all of the remaining tolerable risks to check if any further steps can be taken to reduce the likelihood of occurrence to levels lower than those currently recorded, or the severity of impact reduced if they did occur. If reputation and integrity of your business are important values, then this action is very important.

One way of dealing with the Issues Action Log is to hold a workshop session with all relevant stakeholders, and review each item in detail. Agree what possible action might be taken, what the cost of each action might be, and who might be responsible for taking the action. Establish if any further user, operator, or maintainer training is needed, or if possible redesign or modification is required, etc. Having completed the initial review, and allocated responsibilities for taking action, then you need to set up a schedule for each item, and monitor them to ensure that they are completed by the allocated time. The old adage of *Plan, Do, Review, Repeat*, should be applied!

Decommissioning

A few words should be said about the end of your system's life, and about the decommissioning. In order to understand this, the first thing that needs to be done is the creation of a Decommissioning Plan.

We will assume that you will know when the life of your system is ended; for if the cost of maintenance is excessive, or if it is obsolete because there's some new technology that replaces it, and so on. The Decommissioning Plan should start with a description of the system, and what its inputs and outputs are, together with descriptions of the various subsystems descriptions within the boundary of the existing system. In Chapter 2, we defined the system by creating a model of it, and we can use that to help create the stages for the decommissioning activities.

Using Your System

Having described the system and put a boundary around it, we need to examine the stages in the process. Generically, these are likely to be:

- De-activation
- Removal of components and sub-systems
- Environmental clean-up
- Making good the site

All these activities must be done safely. Power sources, including gas and electricity, need to be disconnected in a safe way, without causing risk to the people doing the work. Compressed gases and volatile substances, which may be harmful to people and the atmosphere, need to be disposed of in accordance with regulations and legislation. Noise and dust need to be considered, and prevented from causing nuisance to neighbours.

Removal of components and sub-systems may require specialist handling equipment. Consideration should be given to the disposal of system elements, whether they have a value, can be recycled, or reused, and so on.

The final step is to determine what will replace your system. From a business point of view, it will ideally be another of your systems, an improved version of the original, or a complete replacement of the functionality.

Enjoying Success

For this closing section, I will say just a few words about enjoying the success that your system is having. Questions to ask yourself include:

- Has your system made you any money?
- What are the rewards, and are you getting them?

These might be financial questions, but there are other benefits, such as enhancing the reputation of you or your company.

You should ask if you benefitted in terms of personal development, or if maybe you have identified a training need to help you do better next time.

There are awards given by a number of professional organisations. The International Council on Systems Engineering (INCOSE) is a good place to start. And certainly, if you want to make a career of systems engineering, then the accreditation body will be helpful in pointing the way. Even if you are comfortable with your achievements to date in your career, then recognition of a job well done by your peers is always a welcome boost to self-esteem.

Finally, sharing your knowledge by writing a paper for, or speaking at, a conference, are straightforward ways of letting people know what you have done, what went well, and what was not so good. Of course, you could always write a book and put it out to the world on the internet!

Using Your System

**There is a website for this book:
www.sensationalsystemsbook.com**

This provides access to a summary of the tips given in this book for successful system delivery. It will also give access to a workbook to help you develop the techniques on which this book is based.

The author, Geoff Miles, is a Chartered Engineer, and runs a coaching, mentoring, and consultancy business. He is also available for teaching and speaking engagements.

**Details can be found at:
www.geoffmilesconsulting.com**

About the Author

Geoff Miles is a Chartered Engineer (CEng), has an honours degree in Mechanical Engineering (BSc Hons), and is a member of the Institution of Mechanical Engineers (MIMechE). Originally, he started out wanting to be a farmer, and although driving a tractor and fixing agricultural machinery was fun, he needed land. His family never owned any, so he borrowed some, and his first crop was a small field of rhubarb.

He got pleasure from rescuing cars in distress, and with guidance from his engineering Uncle Roy, studied for a National Diploma in Agricultural Engineering, which was passed with distinctions. This led to an honours degree in Mechanical Engineering from the University of Aston, in Birmingham. He was sponsored by Metro Cammell, train makers to the world, and after graduation, was appointed as Metro Cammell's man in Hong Kong, for the Mass Transit Railway train fleet.

He won the Sir Stanley Herbert Whitelegg Memorial Travel Scholarship, awarded once every two years by the Institution of Mechanical Engineers, and spent time with train manufacturers and operators in Europe, seeing first-hand the different systems for producing and using different types of train fleets.

Dreams of agricultural success had faded into the dim past, and on return from Hong Kong, he was appointed as Production Control Manager in the Metro Cammell factory. He introduced a new progress control system for the shop floor to improve production efficiency and on-time delivery. One of Metro Cammell's competitors, Alstom, bought the company, and promptly shut it down. He moved on to Strachan & Henshaw, in Bristol, where he introduced an IBM mainframe computer system for controlling production across a range of products, for both the nuclear fuel industry and the Royal Navy.

With recessions looming in both of Strachan & Henshaw's main market areas, he moved back into the rail market area, and was appointed as Contract Manager for the 72-strong fleet of new trains for the extension of the London Docklands Light Railway. Having successfully delivered those, he established the plan for upgrading the signalling system of the fully automated trains, to enable system capacity to be increased, and service reliability improved.
Knowledge of rail infrastructure systems and their interfaces with train systems was increased, and experience in how to make the complexity of interactions more reliable and easier to maintain, was obtained over a series of different assignments as a consultant. Notable amongst the variety of challenging work was getting acceptance to the environmental impacts of an elevated rail system through an urban environment. Noise and vibration was a key issue for that. Another was the task of bringing national rail performance back to previous levels, after rolling contact fatigue was found to have become an epidemic due to changes in rail maintenance system

About the Author

regimes. This resulted in him joining a dedicated *tiger team* of global specialists, to identify the source of the problem, to establish workable solutions for resolution, and implement the solutions throughout the UK rail network.

Rolling stock and depots work at Crossrail, which in its time was the largest civil engineering programme of work in Europe, was able to benefit from his wealth of knowledge and experience in train and rail infrastructure systems, to good effect.

Now, running his own consultancy business, Geoff is able to take a holistic thinking approach to apply project, technical, systems, and design management skills across the spectrum of the project and systems lifecycle, to a wide range of problems, to add real value to client's businesses and to effectively deal with the issues that are of highest concern.

Geoff is able to provide coaching, mentoring, and consulting services across a range of industries and business sizes. He can also take training, speaking, TV, and radio engagements.

Geoff Miles
July 2018
www.geoffmilesconsulting.com

Testimonials

Organisations have become increasingly dependent on electronic delivery of services to meet customer needs. They need high-quality IT Services matched to business needs and user requirements as they evolve. In my work at the CCTA developing PRINCE2™ and "Managing Successful Programmes™", and at the Office of Government Commerce, I learnt first-hand the importance of successful systems and services for all aspects of our lives. I have worked with Geoff in a number of community projects and found that his systems approach to life was immensely helpful in achieving our goals. I'm pleased to commend this book to give you the opportunity to try Geoff's systems techniques for yourself.

Bob Assirati CBE (Hon)FAPM CITP
Director, Sigma Projects UK

During my year as President of the Institution of Civil Engineers, I had the opportunity to travel to a number of different countries around the world and to see engineering development in a range of industries. I realised that there is a common theme as people strive to improve their environment and to sustainably evolve and grow; they need systems to make it all happen within the constraints of safety, time, cost and quality. At the time of UK national rail privatisation, Geoff and I and a small team of other professionals successfully delivered the Government's objective safely, cost effectively and on time using a management system put in place for the purpose. I am confident that you will draw similar inspiration from Geoff's book as you strive to deliver your own life's goals.

Professor Peter Hansford FREng FICE FAPM
University College London

I had the opportunity to work with Geoff Miles on two occasions; firstly when we installed and commissioned the moving block signalling system for the Docklands Light Railway and later when we needed to get the Jubilee Line extension railway systems operational for the VIP launch of the Millennium celebrations. In their way, both were sensational systems events which were successfully delivered. I am delighted to know that Geoff has tried to capture his knowledge and experience in exemplary systems delivery in this book so that others can benefit and hopefully achieve greatness in their own systems fields.

David Waboso CBE, CEng, FREng, FICE, FIRSE HonFAPM
Managing Director, Group Digital Railway
Network Rail

www.ingramcontent.com/pod-product-compliance
Lightning Source LLC
Chambersburg PA
CBHW071537220526
45469CB00003B/819